Transgenic Microalgae as Green Cell Factories

ADVANCES IN EXPERIMENTAL MEDICINE AND BIOLOGY

Recent Volumes in this Series

Volume 60
BREAST CANCER CHEMOSENSITIVITY
Edited by Dihua Yu and Mien-Chie Hung

Volume 609
HOT TOPICS IN INFECTION AND IMMUNITY IN CHILDREN VI
Edited by Adam Finn and Andrew J. Pollard

Volume 610
TARGET THERAPIES IN CANCER
Edited by Francesco Colotta and Alberto Mantovani

Volume 611
PEPTIDES FOR YOUTH
Edited by Susan Del Valle, Emanuel Escher, and William D. Lubell

Volume 612
RELAXIN AND RELATED PEPTIDES
Edited by Alexander I. Agoulnik

Volume 613
RECENT ADVANCES INTO RETINAL DEGENERATION
Edited by Joe G. Hollyfield, Matthew M. LaVail, and Robert E. Anderson

Volume 614
OXYGEN TRANSPORT TO TISSUE XXIX
Edited by Kyung A. Kang

Volume 615
PROGRAMMED CELL DEATH IN CANCER PROGRESSION AND THERAPY
Edited by Roya Khosravi-Far and Eileen White

Volume 616
TRANSGENIC MICROALGAE AS GREEN CELL FACTORIES
Edited by Rosa León, Aurora Gaván, and Emilio Fernández

Transgenic Microalgae as Green Cell Factories

Edited by

Rosa León, Ph.D.
Departamento de Química y Ciencia de Materiales, Facultad de Ciencias Experimentales, Universidad de Huelva, Huelva, Spain

Aurora Galván, Ph.D.
Departamento de Bioquímica y Biología Molecular, Campus de Rabanales, Universidad de Córdoba, Córdoba, Spain

Emilio Fernández, Ph.D.
Departamanto de Bioquímica y Biología Molecular, Campus de Rabanales, Universidad de Córdoba, Córdoba, Spain

Springer Science+Business Media, LCC
Landes Bioscience

Springer Science+Business Media, LLC
Landes Bioscience

Printed in the U.S.A.

Springer Science+Business Media, LLC, 233 Spring Street, New York, New York 10013, U.S.A.
http://www.springer.com

Please address all inquiries to the Publisher:
Landes Bioscience, 1002 West Avenue, 2nd Floor, Austin, Texas, U.S.A. 78701
Phone: 512/ 637 6050; FAX: 512/ 637 6079
http://www.landesbioscience.com

Transgenic Microalgae as Green Cell Factories, edited by Rosa León, Aurora Gaván and Emilio Fernández. Landes Bioscience / Springer Science+Business Media, LLC dual imprint / Springer series: Advances in Experimental Medicine and Biology

ISBN: 978-0-387-75531-1

Library of Congress Cataloging-in-Publication Data

Library of Congress Cataloging-in-Publication Data

Transgenic microalgae as green cell factories / edited by Rosa León, Aurora Gaván, Emilio Fernández.
p. ; cm.
Includes bibliographical references.
ISBN 978-0-387-75531-1
1. Microalgae--Physiology. 2. Microalgae--Biotechnology. I. León, Rosa, Ph. D. II. Gaván, Aurora. III. Fernández, Emilio.
[DNLM: 1. Algae--physiology. 2. Biological Factors--biosynthesis. 3. Bioreactors. 4. Biotechnology--methods. 5. Transformation, Genetic. QK 568.M52 T772 2007]
QK568.M52T73 2007
579.8--dc22

2007035912

PARTICIPANTS

Pio Colepicolo
Department of Biochemistry
University of Sao Paulo
Sao Paulo
Brazil

Konrad Dabrowski
Department of Natural Resources
Ohio State University
Columbus, Ohio
U.S.A.

Almut Eckert
Kompetenzzentrum
für Fluoreszente Bioanalytik
Universität Regensburg
Regensburg
Germany

Vanessa Falcao
Department of Biochemistry
University of Sao Paulo
Sao Paulo
Brazil

Emilio Fernández
Departamanto de Bioquímica
y Biología Molecular
Campus de Rabanales
Universidad de Córdoba
Córdoba
Spain

Samuel P. Fletcher
Department of Cell Biology
The Skaggs Institute
for Chemical Biology
The Scripps Research Institute
La Jolla, California
U.S.A.

Markus Fuhrmann
Sloning BioTechnology
Puchheim
Germany

Aurora Galván
Departamento de Bioquímica
y Biología Molecular
Campus de Rabanales
Universidad de Córdoba
Córdoba
Spain

Maria L. Ghirardi
National Renewable
Energy Laboratory
Basic Science Center
Golden, Colorado
U.S.A.

David González-Ballester
Departamento de Bioquímica
y Biología Molecular
Campus de Rabanales
Universidad de Córdoba
Córdoba
Spain
and
Department of Plant Biology
The Carnegie Institution
of Washington
University of Stanford
Stanford, California
U.S.A.

Christoph Griesbeck
Kompetenzzentrum
für Fluoreszente Bioanalytik
Universität Regensburg
Regensburg
Germany

Arthur R. Grossman
Department of Plant Biology
The Carnegie Institution
Stanford, California
U.S.A.

Markus Heitzer
Kompetenzzentrum
für Fluoreszente Bioanalytik
Universität Regensburg
Regensburg
Germany

Peter Kroth
Fachbereich Biologie
Universität Konstanz
Konstanz
Germany

Rosa León
Departamento de Química
y Ciencia de Materiales
Facultad de Ciencias Experimentales
Universidad de Huelva
Huelva
Spain

Stephen P. Mayfield
Department of Cell Biology
The Skaggs Institute
for Chemical Biology
The Scripps Research Institute
La Jolla, California
U.S.A.

Anastasios Melis
Department of Plant
and Microbial Biology
University of California
Berkeley, California
U.S.A.

Machiko Muto
Department of Cell Biology
The Skaggs Institute
for Chemical Biology
The Scripps Research Institute
La Jolla, California
U.S.A.

Saul Purton
Algal Research Group
Department of Biology
University College London
London
U.K.

Sathish Rajamani
Biophysics Program
Ohio State University
Columbus, Ohio
U.S.A.

Richard Sayre
Biophysics Program
and
Department
of Plant Cellular and Molecular Biology
Ohio State University
Columbus, Ohio
U.S.A.

Michael Seibert
National Renewable Energy Laboratory
Basic Science Center
Golden, Colorado
U.S.A.

Surasak Siripornadulsil
Department of Microbiology
Khon Kaen University
Khon Kaen
Thailand

Moacir Torres
Department of Biochemistry
University of Sao Paulo
Sao Paulo
Brazil

Agustín Vioque
Instituto de Bioquímica Vegetal y Fotosíntesis
Universidad de Sevilla-CSIC
Sevilla
Spain

CONTENTS

PREFACE

Microalgae have been largely cultured and commercialised as food and feed additives, and their potential as source of high-added value compounds is well known. But, in contrast to the large number of genetically modified bacteria, yeast and even higher plants, only a few species of microalgae have been genetically transformed with efficiency. Initial difficulties in the expression of foreign genes in microalgae have been progressively overcome, and powerful molecular tools for their genetic engineering are now on hand. A considerable collection of promoters and selectable marker genes and an increasing number of genomic or cDNA sequences have become available in recent years. More work is needed to transform new species of microalgae, especially those that have commercial value, so that it would be possible to increase the productivity of traditional compounds or synthesize novel ones. Silencing transgenes remains an important limitation for stable expression of foreign genes. This problem is not unique to microalgae since it has also been observed in plants, animals and fungi. A better understanding of the mechanisms that control the regulation of gene expression in eukaryotes is therefore needed.

In this book a group of outstanding researchers working on different areas of microalgae biotechnology offer a global vision of the genetic manipulation of microalgae and their applications. In Chapter 1 the transformation methods and some of the problems and achievements in the genetic manipulation of microalgae, mostly related to the chlorophyte *Chlamydomonas*, are reviewed from the first successful experiments to date. Further chapters provide an ample view of the genetic manipulation of cyanobacteria (Chapter 2) and diatoms (Chapter 3) are also included. The transformation of chloroplasts is an interesting alternative to the nuclear transformation that allows gene integration by homologous recombination and is studied in detail in Chapter 4. To complete the fundamental aspects of the genetic manipulation of microalgae, the influence of codon usage on the expression of heterologous proteins is analyzed in Chapter 5. Bias in codon usage is one of the main limitations for expression of foreign genes in microalgae that has stimulated the search for creative solutions, such as the synthesis of genes adapted to host codon usage.

The book includes chapters focussing on biotechnological applications of transgenic microalgae. The enormous amount information generated by the genomic or cDNA sequencing projects and its implications on the development of new biotechnological applications of microalgae are analysed in Chapter 6. The non-homologous recombination in eukaryotic microalgae has made difficult the directed knockout of selected genes, but in turn it has provided an excellent method for mutagenesis and for the construction of tagged collections of mutants (Chapter 7). Pioneering works about the use of transgenic microalgae as efficient cell factories for the expression of recombinant proteins (Chapter 8), bioremediation (Chapter 9), production of hydrogen (Chapter 10), and recombinant vaccines (Chapter 11) are also included.

We are grateful to the contributing authors for their excellent work. They reviewed the published research work in their area of expertise and shared their unpublished results to bring the chapters up to date. Editing this book on the back of other research and academic duties has been demanding of our time, and we thank family and friends for their tolerance.

Finally, we do hope that the book will prove useful for algologists, biotechnologists, and researchers interested in green cell factories.

Rosa León, Ph.D.
Aurora Galván, Ph.D.
Emilio Fernández, Ph.D.

ABBREVIATIONS

BBS	Bardet-Biedl syndrome
BKD	bacterial kidney disease
Ble	phleomycin resistance marker
C4	carbon fixation through 4 carbon intermediates
CCM	carbon concentrating mechanism
CCMP	Center for Culture of Marine Phytoplankton
cDNA	copy DNA
CFP	cyan fluorescent protein
Chl	chlorophyll
Ci	inorganic carbon
CIA5	inorganic carbon activator protein
CMY	heavy metal biosensor
CP57	chloroplast expressed p57 antigen
E22	nuclear expressed membrane fusion protein with C-terminal, 14 amino acid, p57 epitope
ELIP	early light inducible protein
EST	expressed sequence tag
ESV	*Ectocarpus siliculosus* virus
EXAFS	extended X-ray absorption fine-structure spectroscopy
Fd	ferredoxin
Fo	minimal chlorophyll fluorescence yield (observed in very low light and primarily from photosystem II)
FRET	fluorescence resonance energy transfer
GAPC	gut-associated phagocytotic cells
gDW	grams dry weight
GFP	green fluorescent protein
GPA	glycine, proline, alanine-rich protein
HLIP	high light inducible protein
JGI	Joint Genome Institute
kGDW	kilogram dry weight
LHC	light harvesting complex
LHCA	light harvesting protein A (photosystem I)
LHCB	light harvesting protein B (photosystem II)
Lor	loroxanthin (when lower case and italicized it indicates a mutant in loroxanthin synthesis)
Mb	megabasepairs
MT	metallothionein, class II
NP57	nuclear expressed p57 protein
Npq	nonphotochemical quenching (when lower case and italicized it indicates a mutant strain)
NUPT	nuclear integrant of plastid DNA
P	phosphorus
P5CS	pyrroline-5-carboxylate synthase
PBCV	*Paramedium bursaria chlorella* virus
PHOT	phototropin
PS	photosystem
PSR	phosphorus starvation response
RNAi	RNA interference
ROS	reactive oxygen species
SAC	sulfur acclimation
SDV	silica deposition vesicle
SEPs	stress enhanced protein
TEM	transmission electron microscopy
UV	ultraviolet
WT	wild-type
YFP	yellow fluorescent protein

Chapter 1

Nuclear Transformation of Eukaryotic Microalgae

Historical Overview, Achievements and Problems

Rosa León* and Emilio Fernández

Abstract

Transformation of microalgae is a first step in their use for biotechnological applications involving foreign protein production or molecular modifications of specific cell metabolic pathways. Since the first reliable achievements of nuclear transformation in *Chlamydomonas*, other eukaryotic microalgae have become transformed with molecular markers that allow a direct selection. Different methods—glass beads, electroporation, particle bombardment, or Agrobacterium—and constructions have been set up in several organisms and successfully used. However, some problems associated with efficiency, integration, or stability of the transgenes still persist and are analysed herein. Though the number of microalgae species successfully transformed is not very high, prospects for transformation of many more are good enough on the basis of what has been achieved so far.

Introduction

Microalgae constitute a highly heterogeneous group of prokaryotic and eukaryotic organisms with a capital ecological importance, accounting for about 50% of the global organic carbon fixation[1] and having an enormous biotechnological potential. Many species have been used for the production of high-added-value compounds of application in feeding, dietetics, cosmetics and fine chemistry industries; and in various processes such as wastewater treatment or biofertilization.[2-4] Furthermore, microalgae are used as model systems for studying fundamental processes such as photosynthesis, flagellar function, photoreception and nutrient acquisition[5,6] and have been proposed as an alternative system for the expression of heterologous proteins, including antibodies and other therapeutic proteins.[7,8]

In contrast to the large number of genetically modified bacteria, yeast and even higher plants, only a few species of eukaryotic microalgae have been successfully transformed with a certain efficiency. The development of molecular tools for efficient and stable genetic manipulation of microalgae is therefore necessary to enhance their potential for engineering their metabolic pathways.

The best studied genetically modified eukaryotic microalga is so far the freshwater chlorophyte *Chlamydomonas reinhartii*, which was first transformed in 1989 by complementation of *nit1* and *arg7* mutations with homologous nitrate reductase[9] and argininosuccinate lyase genes,[10] respectively. Since then a significant number of selectable markers, promoters, and new procedures

*Corresponding Author: Rosa León—Deparamento de Química y Ciencia de Materiales, Facultad de Ciencias Experimentales, Avda, Fuerzas Armadas s/n, Universidad de Huelva, 21007-Huelva, Spain. Email: rleon@uhu.es

Transgenic Microalgae as Green Cell Factories, edited by Rosa León, Aurora Galván and Emilio Fernández. ©2007 Landes Bioscience and Springer Science+Business Media.

Table 1. Summary of the main species of microalgae genetically modified and the transformation method used

Division	Species	Method	Ref.
Dinoflagellates	Amphidinium	Silicon carbide whiskers	37
	Symbiodinium	Silicon carbide whiskers	37
Diatoms	*Phaeodactylum tricornutum*	Bombardment	36,32
	Cyclotella críptica	Bombardment	34
	Navicula saprophila	Bombardment	34
	Cylindrotheca fusiformis	Bombardment	35
	Thalassiosira weissflogii	Bombardment	36
Chlorophyceae	*Chlamydomonas*	Glass beds	39
		Electroporation	42
		Silicon carbide whiskers	41
		Bombardment	10,44
		Agrobacterium	45
	*Chlorella ellipsoidea**	Bombardment	19
		Electroporation	23
	Chlorella saccharophila	Electroporation	20
	*Chlorella vulgaris**	Protoplast transformation	21
	*Haematococcus pluvialis**	Bombardment	26
	Dunaliella salina	Electroporation	29,27
		Bombardment	28,30

* In some cases only transient expression has been observed.

for efficient introduction of DNA into microalgal nucleus have been developed and transformation efficiency has dramatically increased. Nevertheless the number of transformed species has timidly increased to about a dozen of new strains (Table 1).

Excellent reviews on different aspects of genetic transformation of microalgae,[11-13] some focused exclusively on *Chlamydomonas*,[14,-16] have been previously published. Here, we will review the main methods and strategies presently used for nuclear transformation of eukaryotic microalgae, including those species that have been successfully transformed and discuss the main difficulties associated with stable expression of transgenes. The transformation of chloroplast and cyanobacteria has specific characteristics that are treated in Chapters 4 and 2, respectively.

Microalgae Groups Transformed

Microalgae are phylogenetically very heterogeneous. Until now, there are reports of stable nuclear transformations in three eukaryotic microalgal groups: Chlorophytes, diatoms, and dinoflagellates (Table 1).

Chlorophytes

The freshwater alga *Chlamydomonas* is the first and best studied transformed chlorophyte[14,15,16] that has already become a powerful model system for molecular studies[5,6,17,18] due to its easy manipulation techniques and the availability of bioinformatic tools, such as an EST database (http://www.chlamy.org) and a draft of the complete genome sequence (http://genome.jgi-psf.org/Chlre3/Chlre3.home.html). A large variety of transformation methods and constructions have been designed for this microalga (see Table 2) and several biotechnological processes involving transgenic *Chlamydomonas* have been described. Production of H_2 (see Chapter 10), recombinant vaccines (Chapter 11) and bioremediation (Chapter 9) are some examples.

Table 2. Examples of marker and reporter genes used in microalgal constructs

Gene	Description	Gene Source	Ref.
aadA	Adenylyl transferase (resistance to spectinomycin)	*Eubacteria*	74
als	Acetolactate synthase (resistance to sulfonylurea herbicides)	*Chlamydomonas*	70
aphVIII	Aminoglycoside 3'phosphotransferase (resistance to paramomycin)	*Streptomyces rimosus*	69
ars	Arylsulphatase	*Chlamydomonas*	59
ble	Bleomycin binding protein (resistance to zeocin)	*Streptoalloteichus hindustanus*	83
cat	Chloramphenicol acetyltransferase (resistance to chloramphenicol)	Transposon Tn9	66
cry1-1	Ribosomal protein S14	*Chlamydomonas*	52
ε-frustulin	Calcium binding glycoprotein	*Navicula pelliculosa*	35
gfp[c]	Modified green fluorescent protein	Adapted to *Chlamydomonas* codon usage	57
gfp[e]	Modified green fluorescent protein	Adapted to human codon usage	33
glut1	Glucose transporter	Human	84
gus	β-glucoronidase	*Escherichia coli*	37
hpt	Hygromycin B phosphotransferase	*Escherichia coli*	37
hup1	Hexose transporter	*Chlorella kessleri*	84
luc	Luciferase	*Horatia parvula* (firefly)	51
nat	Nourseothricin resistance	*Streptomyces noursei*	33
nptII	Neomycin phosphotransferase II (resistance to G418)	*Escherichia coli*	64
oee-1	Oxygen evolving enhancer protein	*Chlamydomonas*	85
sat-1	Nourseothricin resistance	*Escherichia coli*	33

The first successful introduction and expression of foreign DNA in *Chlorella* cells reports on the transient expression of firefly luciferase (*Luc*) in protoplasts of *C. ellipsoidea*[19] and of β-glucuronidase (*Gus*) gene in *C. saccharophila* by electroporation.[20] The expression of the human growth hormone (*hGH)* under the control of different promoters including a *Chlorella* virus promoter, the RbcS2 promoter from *Chlamydomonas reinhardtii* and the CaMV35S promoter was also achieved in *C. vulgaris* and *Chlorella sorokiniana* protoplasts.[21] Although expression was not stable for more than a couple of months, these authors were able to induce excretion of the heterologous protein by inserting an extracellular secretion signal sequence between the promoter and the *hGH* gene. The first evidence of stable transformation of *Chlorella* was the complementation of nitrate reductase deficient mutants of *C. sorokiniana* with a homologous gene of nitrate reductase from *C. vulgaris* by microprojectile bombardment that was reported as an event of homologous recombination.[22] Difficulties for getting stable transgene integration and expression in *Chlorella* transformants will require further studies to ascertain the relevance of homologous integration. Anyway, expression of commercially interesting genes such as the heterologous rabbit neutrophil peptide-1 (NP-1),[23] the flounder growth hormone,[24] and a bio-insecticide-acting peptide hormone from mosquito ovaries[25] has been reported in *Chlorella* though stability of the expression of these genes is not clearly stated.

Other chlorophytes with important economic value, such as *Dunaliella* or *Haematococcus*, have been for a long time refractory to any type of genetic manipulation, and only very recently some promising reports of successful transformation have arisen. Transient expression of β-galactosidase in *Haematococcus pluvialis* by bombarding the *LacZ* gene driven by the SV40 promoter,[26] and integration into *Dunaliella* genome of foreign genes by electroporation[27] or micro-particle bombardment[28] were described. However, stable foreign gene expression has only recently been announced.[29,30] The lack of specific promoters for these species is the main limitation for the development of efficient and stable transformation systems. The flanking regions of *Dunaliella rbcS2* gene were used to drive the expression of marker genes in *C. reinhardtii*,[31] opening the way for the efficient genetic manipulation of this halophilic microalga, which is the main source of natural beta-carotene.

Diatoms

Several strains in this group have important biotechnological applications derived from their silicean cell wall and their use in aquaculture and as source of particular oils[2] (Apt and Behrens, 1999). Their genetic manipulation and biotechnological applications are presented in Chapter 3. *Phaedactylum tricornutum*,[32,33] *Cyclotella criptica*,[34] *Navicula saprophila*,[34] *Cylindrotheca fusiformis*,[35] and *Thalassiosira weissflogii*[36] are some of the strains successfully transformed.

Dinoflagellates

Dinoflagellates or pyrrophyta are a group of unicellular eukaryotic alveolar algae that constitute an important part of marine phytoplankton. Some species of this division produce luminescent compounds; others produce compounds that are toxic to vertebrates including humans. Species of this group have typical eukaryotic cytoplasmatic features but the characteristics of their nucleus are relatively unique. They lack histones and combine typical prokaryote characteristics, such as permanently condensed chromosomes, with eukaryotic features, such as the presence of nucleolus and introns. The strains *Amphidinium* and *Symbiodinium* have been successfully transformed.[37] In both cases intact walled cells were transformed by agitation in the presence of silicon carbide whiskers. The *Gus* gene, fused to neomycin phosphotransferase (*nptII*) or hygromycin phosphotransferase (*hpt*) genes, that confer resistance to kanamycin or hygromycin were expressed under the control of either *Agrobacterium* p1'2'promoter or cauliflower mosaic virus 35S promoter. The unique nuclear characteristics of these microalgae can influence their ability to express genes driven by heterologous promoters. Although prolonged selection is required, dinoflagellate genera *Amphidinium* and *Symbiodinium* might be considered as one of the most accessible groups of algae for genetic manipulation.

Others

The scope of this review does not include multicelullar agae, although it should be noted that efficient transformation systems have been developed in species such as *Volvox cateri*,[38] the carrageenan-producing rhodophyte *Kappaphycus alvarezii* or the green macroalga *Ulva lactuca*.[11]

Methods for Microalgae Transformation

A variety of alternative approaches for gene transfer to eukaryotic microalgal nucleus have been developed. The basis of the traditional methods used to transform microalgae is to cause, by different means, a temporal permeabilization of the cell membrane, enabling DNA molecules to enter the cell while preserving viability.

Agitation in the Presence of Glass Beads or Silicon Carbide Whiskers

In the glass beads method, permeabilization of the cells is obtained by vortexing in the presence of DNA, glass beads and PEG.[39] It was originally developed for yeasts and has been successfully used for transformation of cell-wall deficient mutants or wild-type cells of *Chlamydomonas* following enzymatic degradation of the cell wall. No reports about the efficiency of this method for transforming other species exist, but the simplicity and efficiency of this procedure make it very interesting to investigate whether it can be applied to other species of microalgae. The main advantage of this method is that it does not require specialized equipment. Though the main drawback is considered to be the requirement of cell wall-less strains as hosts, low efficient transformations are achieved even with intact walled *Chlamydomonas* cells.[40]

Dunahay[41] (1993) described a similar method using agitation with silicon carbide whiskers for *Chlamydomonas* transformation without removing of the cell wall. This method has also been successfully used for the transformation of the dinoflagellates, *Amphidinium* and *Symbiodinium*.[37]

Electroporation

In electroporation, transient holes in the cell membrane are formed when pulses of an electric current are applied. The introduction of genes by electroporation has been widely carried out in animal, plant and bacterial cells. Stable transformants of both wall-less and walled strains of the microalga *Chlamydomonas reinhardtii*,[42] *Chlorella ellipsoidea*,[23] and *Dunaliella salina*[29] have been obtained by this method. Temperature, osmolarity, electric conditions, field strength ($kVcm^{-1}$), time of discharge, and DNA concentrations have to be carefully optimised to obtain high transformation efficiencies. High efficiency of transformation is achieved in *Chlamydomonas* to 2×10^{-5} transformants per cell and µg of DNA, about two orders of magnitude higher than that obtained with the standard glass beads method to introduce exogenous DNA.[42]

Microparticle Bombardment

In the microparticle bombardment or biolistic method the DNA is literally shot into the host cells with a "gene gun". Small gold or tungsted particles (0.5-1.5 µm) are coated by DNA and accelerated (at about $500m\ s^{-1}$) by an Helium driven gun into the target cells. It has been mainly applied for transformation of plant cells and tissues[43] but it has also been successfully used for transformation of other biological systems as diverse as animal cells tissues, fungi, subcelullar organella, bacteria and algae. The technique has demonstrated to be particularly successful for transformation of microalgae that are refractory to other transformation methods, such as diatoms[32,34] or walled clorophytes.[44]

Agrobacterium tumefaciens

Agrobacterium tumefaciens is a common tool in plant genetic transformation because it has evolved the unique capacity to transfer a piece of its own DNA (the T-DNA) from the Ti plasmid into the nuclear genome of plant cells. This represents an example of naturally occurring trans-kingdom transfer of genetic material and has enabled this bacterium to be used as a tool to integrate selected genes in plant chromosomes. The method was initially applied to dicotyledonous plants, which are the natural host for this pathogenic soil bacteria, but it was soon used for genetic manipulation of monocotyledonous plants, including cereals, and fungi. The stable genetic transformation of the unicellular microalgae *C. reinhardtii* by *Agrobacterium* has recently been reported.[45] These authors cocultivated *Chlamydomonas* and *Agrobacterium* cells carrying a plasmid which contains the hygromycin phosphotransferase *(hpt)* gene and the gene fusion *gfp:uidA* as reporter gene, both driven by the CaMV 35S promoter. Transformation efficiency obtained by this method is 50-fold higher than that of the glass beads transformation. Until now no other reports on microalgae transformation using this method have been published. However the very wide host range of *Agrobacterium*, including gymnosperms and perhaps lower plant phyla, a variety of fungi, and even animal cells, suggests that T-DNA transfer might become a common system to deliver genes in algae.[46]

Other Methods

Other approaches, such as the use of recombinant viruses as vehicles to carry foreign DNA into algae, the utilization of transposons elements or microinjection are being studied. There are early reports of transformation by microinjection of the giant (2-4cm) algae *Acetabularia*[47] which is unicellular for most of its lifecycle. But application of this method to usual microscopic microalgae has obvious size limitations.

Several viruses infecting *Chlorella* and brown algae are being considered for developing both cloning vehicles and expression systems. A *Chlorella* virus promoter sequence was successfully used by Hawkins and Nakamura[21] (1999) for transient expression of a gene encoding for human growth hormone (*hGH*). Reciprocally, a *Chlorella* virus adenine methyltransferase gene promoter functions as a strong promoter in plants, and is significantly stronger than the cauliflower mosaic virus 35S promoter.[48] The large size (150-330 kpb) of the algal viral genomes may permit insertion of foreign genes and provide mechanisms for directing their insertion into algal chromosomes,[49,50] but more extensive studies about eukaryotic algal viruses are necessary before this exciting approach became a reality.

Characteristics of the Transformation Process

In contrast to prokaryotes or organelles, foreign DNA preferently integrates at apparently random locations of the eukaryotic nuclear genome by nonhomologous recombination, difficulting the deletion or repairing of the selected genes of interest.[51] This obstacle has been exploited as a method for insertional mutagenesis, whose fundamentals and applications are addressed in Chapter 7. Homologous recombination in nuclear genes was observed at low frequency in *Chlamydomonas*,[52,53] and recently this frequency has been increased by two orders of magnitude by using single stranded DNA.[54]

The number of copies of the introduced DNA varies from one to several depending on the DNA concentration and the transformation technique.[5] It has been described in *Chlamydomonas* that by using low concentrations of DNA (about 100 ng or less per 10^8 cells) most integrations occur with a single copy of marker DNA.[55] In transformants bearing multiple copies of the marker, these are often integrated as concatamers of transforming DNA.[44] This is expected from the amount of DNA (1 μg and higher) routinely used in transformations. The transformation with linearized DNA seems to be more efficient than with circular DNA, and intregration events are more predictable. Nevertheless, it is important that the chosen cleavage site is not very close to the 5′ or 3′ UTR regions of the introduced gene to preserve its integrity.[56]

One of the difficulties found to express heterologous genes in microalgae is the selection of transformants. Cotransformation with two different plasmids, one of which contains an easily selectable marker, can solve the problem, although frequency of expression of both genes can be low (10-50 %). This efficiency increases dramatically if the two genes are within the same construction.[57]

DNA Constructions Used in Transformation

Availability of marker and reporter genes is of key importance for selection of the transformed microalgae. For several years, the only adequate selectable markers available for microalgae transformation were based on homologous genes (*arg7, nia1=nit1, oee1* or *atpC*) that allowed complementation of specific mutants. But this strategy is not applicable to wild type or diploid microalgae (i.e., diatoms) owing to the difficulties to generate mutants in which both alleles of a gene are defective. Nowadays, a large collection of reporter genes and selectable markers are available, some of which are summarized in Table 2.

The most powerful selectable markers are those that confer resistance against antibiotics such as bleomycine (*Ble*), streptomycin (*aadA*) and paramoycin (*aphVIII*) or to herbicides, such as the gene Acetolactate synthase (*als*) that confers resistance to sulphonylurea herbicides. Some reporter genes as *Gus* (β-Glucuronidase) and adapted versions of *GFP* could also be used to test protein subcellular localization when fused to the desired protein.[16,58] The *Luc* gene, that codes for luciferase of *Renilla reniformis* adapted to *C. reihardtii* nuclear codon usage, expressed under the control of different promoters has been used for "in vivo" monitoring nuclear gene expression.[51] The *ars* gene also provides an easy and rapid assay for promoter activity studies, as shown in *Chlamydomonas*.[59,60]

The choice of highly active promoters to drive the expression of marker and reporter genes is a critical step in the development of an efficient transformation system. Strong constitutive

Table 3. Selected constructs used for microalgal transformation

Microalgal Host	Promoter	Fused Gene	Ref.
Chlamydomonas	*nia1* (Nitrate reductase from *C. reinhardtii*)	*ars*	59,60,63
"	*rbcS2* (Rubisco small subunit from *C. reihardtii*)	*gfp+ble*	57
"	"	*aphVIII*	67
"	"	*ble*	83
"	"	*aadA*	74
"	"	*als*	70
"	*hsp70A* (Heat shock protein from *C. reinhardtii*)	*ars*	61
"	*psaD* (Photosystem I complex protein from *C. reinhardtii*)	*ble*	71
"	*cop* (Chlamyopsin from *C. reinhardii*)	*gfp+cop*	57
"	*nos* (Nopaline synthase from *A. tumefaciens*)	*nptII+nia1*	45
"	*CaMV35S* (Cauliflower mosaic virus 35S)	*Cat*	66
Chlorella	*ubi1-Ω* (ubiquitin-Ω from *Zea mais*)	*Gus*	23
Haematococcus	*SV40* (simian virus 40)	*LacZ*	26
Dunaliella	*ubi1-W* (ubiquitin-Ω from *Zea mais)*	*Gus*	29
Phaeodactylum tricornutum	*fcp* (Fucoxanthin chlorophyll-a or -c binding protein)	*Ble*	32
"	"	*glut-1/hup1*	84
"	"	*ble/nptII/gfp/gus*	33
"	"	*ε-frustulin*	35
"	"	*ble* and *luc*	36
Amphidinium and	*nos* (Nopaline synthase from *A. tumefaciens*)	*NptII*	37
Symbiodinium	*p1'2'* (from *A. tumefaciens*)	*hpt*	37

homologous promoters, such as the *rbcS2* gene promoter of *C. reinhardtii* or the *fcp*A and B promoters of *P. tricornutum* are widely used to drive the expression of heterologous proteins. The upstream fusion of a *hsp70A* promoter fragment enhances activity of neighbouring promoters[61] and decreases the probability of transcriptional silencing,[62] which is an important problem in the expression of transgenes. The use of inducible promoters like the nitrate reductase gene promoter of *Chlamydomonas,*[59,60,63] or the promoter derived from the nitrate reductase gene of the diatom *Cylindrotheca fusiformis,*[64] provides a greater control in the expression. Description of selected promoters used to express heterologous proteins in microalgae is shown in Table 3.

Attempts to express genes fused to heterologous promoters in diatoms were unsuccessful.[32] In Chlorophytes, transformation frequency with heterologous constructs was low[65] or expression was unstable.[28,66] The only report of stable transformation of microalgae with heterologous genes under the control of heterologous promoters is the expression of *Gus* driven by either the cauliflower mosaic virus 35S promoter or the p1'2' *Agrobacterium* promoter in dinoflagellates *Amphidinium* and *Symbiodinium.*[37] The unique nuclear characteristics of these microalgae can influence their ability to express genes under the control of heterologous promoters. A certain frequency of transformation has been achieved with promoter-less heterologous genes in *Chlamydomonas,* probably as a result of in vivo gene-fusion with endogenous promoter regions.[56,67] Identification of promoters ('promoter trapping') was achieved in *Chlamydomonas* with arg7 as a selectable marker, and a promoter-less reporter gene (*rsp3*), which was expressed in 2-3% of the transformants.[68] However, it is generally accepted that, in chlorophytes and diatoms, stable expression of heterologous genes can only be optimally achieved when adequate homologous promoters or promoters from very close species are included.[31,34]

The presence of introns was also shown to be a factor for efficient expression. *Ble* and *aphVIII* genes, under *rbcS2* promoter, increased their expression by including the first intron of *rbcS2*.[69] Expression of the *als* gene under *rbcS2* promoter was also optimal when all *als* introns were present.[70] However, introns are not absolutely required for gene expression as shown with the cDNA of *arg7*.[56] Expression vectors having promoter and *cis* elements from the highly expressed gene *psaD*, lacking introns, are useful to allow high-expression of endogenous and exogenous cDNAs.[71]

Difficulties for Stable Expression of the Transgenes

A great number of regulatory elements and transcriptional or post transcriptional events can influence on the expression level of the transgenes and on their stability. Expression of an exogenous gene can be very low or null, even though all the elements required for optimal transcription and translation -promoters, introns and other regulatory regions- have been included in the chimeric gene construction. Futhermore, when transgenic algal clones are not maintained under selection conditions, expression of the exogenous genes might be suppressed. This need can be avoided in *Chlamydomonas* by freezing transgenic cells in liquid nitrogen for which a protocol has been established (R. Sayre at http://www.chlamy.org/methods/freezing.html and Crutchfield et al[72]) and optimized (http://www.uco.es/organiza/departamentos/bioquimica-biol-mol/english/files/Chlamy_cryopreservation.pdf).

Difficulties for foreign gene expression in microalgae can be due to the lack of adequate regulatory sequences, to positional effects, biased codon usage—that is analysed in Chapter 5—incorrect polyadenylation, inappropriate nuclear transport, unstability of the mRNA, or gene silencing. Gene silencing has been attributed to a variety of epigenetic mechanisms similar to those observed in plants and other eukaryotic cells[73] and is thought to be related to the control of development and to the response of a cell to viruses, transposable elements, or transgenes.[74,75] Epigenetic processes are defined as heritable changes in gene expression without modification of DNA sequences. These changes might occur at the level of transcription or post-transcription by mechanisms that are not yet fully understood. Several elements involved in the silencing of transgenes in *Chlamydomonas* have recently been isolated.[76] Transcriptional gene silencing can occur through an altered chromatin structure that is associated with cytosine methylation of the promoter regions. However, single-copy transgenes can be transcriptionally silenced without detectable cytosine methylation of introduced DNA.[75,77,78] Post-transcriptional transgene silencing can take place by the RNA-mediated process called interference RNA (RNAi). RNAi forms RNA-induced silencing complexes that promote RNA degradation.[75] It is worth noting that mechanistic connections between epigenetic transcriptional silencing and DNA double-strand-break repair have recently been proposed.[78] This disadvantage for gene expression can result in the useful technique of RNA interference (RNAi) that has already helped to the functional characterization of a good number of genes[79] (see Chapter 7).

In addition, chromatin states are determined by Lys methylation of histones mostly mediated by SET domain-containing proteins and depending on the methylated histone residue, the degree of methylation, and the chromatin context, transcriptional repression or activation occurs.[80,81] It has been recently shown that Mut11p, required for transcriptional silencing in *Chlamydomonas reinhardtii*,[78] interacts with conserved components of histone methyltransferase machineries, and proposed that the functional differentiation between dimethylated and monomethylated H3 lysine4 operates as an epigenetic mark for repressed euchromatin.[82] This intricate circuit of methylations might provide methodologies and tools to activate or inhibit, as needed, gene expression in different algal systems.

Concluding Remarks

A speedy progress has been attained since the first reliable transformation techniques in the microalga *Chlamydomonas*. Several methodologies has been developed that primarily rely on an efficient and reproducible transformation. Though much progress has been achieved in other different microalgae, there is still the need to develop general strategies to introduce DNA stably in any algal system. These may come from understanding more deeply the molecular biology of these systems. Tools like wide-spectrum general promoters, DNA markers, and ssDNA transformation, or *Agrobacterium* might provide future developments. Time will tell us.

References

1. Field CB, Behrenfeld MJ, Randerson JT et al. Primary production of the biosphere: Integrating terrestrial and oceanic components. Science 1998; 281:237-240.
2. Apt KE, Behrens PW. Commercial developments in microalgal biotechnology. J Phycol 1999; 35:215-226.
3. Richmond A. Handbook of microalgal culture. Biotechnology and applied phycology. Oxford: Blackwell Science Ltd., 2004.
4. Grossman AR. Paths toward algal genomics. Plant Physiol 2005; 137:410-427.
5. Harris E. Chlamydomonas as a model organism. Annu Rev Plant Physiol Plant Mol Biol 2001; 52:363-406.
6. In: Rochaix JD, Goldschmidt-Clermont M, Merchant S, eds. The Molecular Biology of Chloroplasts and Mitochondria in Chlamydomonas. Dordtrech: Kluwer Acad Pub., 1998.
7. Franklin SE, Mayfield SP. Prospects for molecular farming in the green alga Chlamydomonas reinhardtii. Curr Opin Plant Biol 2004; 7:159-165.
8. Walker TL, Purton S, Becker DK et al. Microalgae as bioreactors. Plant Cell Rep 2005; 24(11):629-641.
9. Fernández E, Schnell R, Ranum LP et al. Isolation and characterization of the nitrate reductase structural gene of Chlamydomonas reinhardtii. Proc Natl Acad Sci USA 1989; 86(17):6449-6453.
10. Debuchy R, Purton S, Rochaix JD. The argininosuccinate lyase gene of Chlamydomonas reinhardtii: An important tool for nuclear transformation and for correlating the genetic and molecular maps of the ARG7 locus. EMBO J 1989; 8:2803-2809.
11. Stevens DR, Purton S. Genetic engineering of eukaryotic algae: Progress and prospects. J Phycol 1997; 33:713-722.
12. León-Bañares R, González-Ballester D, Galván A et al. Transgenic microalgae as green cell-factories. Trends Biotechnol 2004; 22:45-52.
13. Walker TL, Collet C, Purton S. Algal transgenics in the Genomic era. J Phycol 2005; 41(6):1077-1093.
14. Kindle KL. Nuclear transformation: Technology and Applications. In: Rochaix JD, Goldschmidt-Clermont M, Merchant S, eds. The Molecular Biology of Chloroplasts and Mitochondria in Chlamydomonas. Dordrecht: Kluwer Acad Pub., 1998:41-61.
15. Purton S, Lumbreras V. Recent Advances in Chlamydomonas transgenics. Protist 1998; 149:23-27.
16. Fuhrmann M. Expanding the molecular toolkit for Chlamydomonas reinhardtii-from history to new frontiers. Protist 2002; 153:357-364.
17. Snell WJ, Pan J, Wang Q. Cilia and flagella revealed: From flagellar assembly in Chlamydomonas to human obesity disorders. Cell 2004; 117:693-697.
18. Grossman AR, Harris EE, Hauser C et al. Chlamydomonas reinhardtii at the crossroads of genomics. Eucaryotic Cell 2003; 2(6):1137-1150.
19. Jarvis EE, Brown LM. Transient expression of firefly luciferase in protoplasts of the green alga Chlorella ellipsoidea. Curr Genet 1991; 19:317-321.
20. Maruyama M, Horáková I, Honda H et al. Introduction of foreign DNA into Chlorella saccharophila by electroporation. Biotechnol Techn 1994; 8:821-826.
21. Hawkins R, Nakamura M. Expression of human growth hormone by the eukaryotic alga, Chlorella. Curr Microbiol 1999; 38:335-341.
22. Dawson HN, Burlingame R, Cannons AC. Stable transformation of Chlorella: Rescue of nitrate reductase-deficient mutants with the nitrate reductase gene. Curr Microbiol 1997; 35:L365-362.
23. Chen Y, Wang Y, Sun Y et al. Highly efficient expression of rabbit neutrophil peptide-1 gene in Chlorella ellipsoidea cells. Curr Genet 2001; 39:365-370.
24. Kim DH, Kim YT, Cho JJ et al. Stable integration and functional expression of flounder growth hormone gene in transformed microalga, Chlorella ellipsoidea. Mar Biotechnol 2002; 4:63-73.
25. Borovsky D. Trypsin-modulating oostatic factor: A potential new larvicide for mosquito control. J Exp Biol 2003; 206:3869-3875.
26. Teng C, Qin S, Liu J et al. Transient expression of lacZ in bombarded unicellular green alga Haematococcus pluvialis. J Appl Phycol 2002; 14:495-500.

27. Sun Y, Yang Z, Gao X et al. Expression of foreign genes in Dunaliella by electroporation. Mol Biotechnol 2005; 30(3):185-192.
28. Tan C, Quin S, Zhang Q et al. Establishment of a micro-particle bombardment transformation system for Dunaliella salina. J Microbiol 2005; 43(4):361-365.
29. Geng D, Wang Y, Wang P et al. Stable expression of hepatitis B surface antigen gene in Dunaliella salina. J Appl Phycol 2003; 15:451-456.
30. Lü YM, Jiang GZ, Niu XL et al. Cloning and functional analyses of two carbonic anhydrase genes from Dunaliella salina. Acta Genet Sin 2005; 31:1157-1166.
31. Walker TL, Becker DK, Collet C. Characterisationof the Dunaliella tertiolecta RbcS genes and their promoter activity in Chlamydomonas reinhardtii. Plant Cell Rep 2005; 23:727-735.
32. Apt KE, Kroth-Pancic PG, Grossman AR. Stable nuclear transformation of the diatom Phaeodactylum tricornutum. Molec Gen Genet 1996; 252:572-579.
33. Zaslavskaia LA, Lippmeier JC, Kroth PG et al. Transformation of the diatom Phaeodactylum tricornutum (Bacillariophyceae) with a variety of selectable markers and reporter genes. J Phycol 2000; 36:379-386.
34. Dunahay TG, Jarvis EE, Roessler PG. Genetic transformation of the diatoms Cyclotella cryptica and Navicula saprophila. J Phycol 1995; 31:1004-1012.
35. Fischer H, Robl I, Sumper M et al. Targeting and covalent modification of cell wall and membrane proteins heterologously expressed in the diatom Cylindrotheca fusiformis. J Phycol 1999; 35:113-120.
36. Falciatore A, Casotti R, Leblanc C et al. Transformation of nonselectable reporter genes in Marine Diatoms. Mar Biotechnol 1999; 1:239-251.
37. ten Lohuis MR, Miller DJ. Genetic transformation of dinoflagellates (Amphidinium and Symbiodinium): Expression of GUS in microalgae using heterologous promoter constructs. Plant J 1998; 13:427-435.
38. Schiedlmeier B, Schmitt R, Muller W et al. Nuclear transformation of Volvox carteri. Proc Natl Acad Sci USA 1994; 91:5080-5084.
39. Kindle KL. High-frequency nuclear transformation of Chlamydomonas reinhardtii. Proc Natl Acad Sci USA 1990; 87:1228-1232.
40. Quesada A, Galván A, Fernández E. Identification of nitrate transporters in Chlamydomonas reinhardtii. Plant J 1994; 5:407-419.
41. Dunahay TG. Transformation of Chlamydomonas reinhardtii with silicon carbide whiskers. Biotechniques 1993; 15:452-460.
42. Shimogawara K, Fujiwara S, Grossman A et al. High-efficiency transformation of Chlamydomonas reinhardtii by electroporation. Genetics 1998; 148:1821-1828.
43. Sanford JC, Smith FD, Russell JA. Optimizing the biolistic process for different biological applications. Methods in Enzymology 1993; 217:483-509.
44 Kindle KL, Schnell RA, Fernández E et al. Stable nuclear transformation of Chlamydomonas using the Chlamydomonas gene for nitrate reductase. J Cell Biol 1989; 109(6):2589-601.
45. Kumar SC, Misqitta RW, Reddy VS et al. Genetic transformation of the green alga Chlamydomonas reinhardtii by Agrobacterium tumefaciens. Plant Sci 2004; 166:731-738.
46. Gelvin SB. Agrobacterium-mediated plant transformation: The biology behind the gene jockey tool. Microbiol Mol Biol Rev 2003; 67:16-37.
47. Langridge P, Brown JWS, Pintor-Toro JA et al. Expresión of zein genes in Acetabulara mediterranea. Eur J Cell Biol 1985; 39:257-64.
48. Mitra A, Higgins DW. The Chlorella virus adenine methyltransferase gene promoter is a strong promoter in plants. Plant Mol Biol 1994; 26:85-93.
49. Henry EC, Meints RH. Recombinant viruses as transformation vectors of marine microalgae. J Appl Phycol 1994; 6:247-253.
50. Etten V, Meints RH. Giant viruses infecting algae. Annu Rev Microbiol 1999; 53:447-494.
51. Fuhrmann M, Hausherr A, Ferbitz L et al. Monitoring dynamic expression of nuclear genes in Chlamydomonas reinhardtii by using a sythetic luciferase reporter gene. Plant Mol Biol 2004; 55:869-881.
52. Nelson JA, Lefebvre PA. Targeted disruption of the NIT8 gene in Chlamydomonas reinhardtii. Mol Cell Biol 1995; 15:5762-5769.
53. Sodeinde OA, Kindle KL. Homologous recombination in the nuclear genome of Chlamydomonas reinhardtii. Proc Natl Acad Sci USA 1993; 90(19):9199-9203.
54. Zorin B, Hegemann P, Sizova I. Nuclear-gene targeting by using single-stranded DNA avoids illegitimate DNA integration in Chlamydomonas reinhardtii. Eukaryotic cell 2005; 4(7):1264-1272.
55. González-Ballester D, de Montaigu A, Higuera JJ et al. Functional genomics of the regulation of the nitrate assimilation pathway in Chlamydomonas. Plant Physiol 2005; 137(2):522-533.
56. Auchincloss AH, Loroch AI, Rochaix JD. The argininosuccinate lyase gene of Chlamydomonas reinhardtii: Cloning of the cDNA and its characterization as a selectable shuttle marker. Mol Gen Genet 1999; 261(1):21-30.

57. Fuhrmann M, Oertel W, Hegemann P. A synthetic gene coding for the green fluorescent protein (GFP) is a versatile reporter in Chlamydomonas reinhardtii. Plant J 1999; 19(3):353-361.
58. Apt KE, Zaslavkaia LA, Lippmeier JC et al. In vivo characterization of diatom multipartite plastid targeting signals. J Cell Sci 2002; 115:4061-4069.
59. Ohresser M, Matagne RF, Loppes R. Expression of the arylsulphatase reporter gene under the control of the nit1 promoter in Chlamydomonas reinhardtii. Curr Genet 1997; 31(3):264-271.
60. Llamas A, Igeno MI, Galvan A et al. Nitrate signalling on the nitrate reductase gene promoter depends directly on the activity of the nitrate transport systems in Chlamydomonas. Plant J 2002; 30(3):261-271.
61. Schroda M, Blocker D, Beck CF. The HSP70A promoter as a tool for the improved expression of transgenes in Chlamydomonas. Plant J 2000; 21(2):121-131.
62. Schroda M, Beck CF, Vallon O. Sequence elements within an HSP70 promoter counteract transcriptional transgene silencing in Chlamydomonas. Plant J 2002; 31(4):445-455.
63. Koblenz B, Lechtreck KF. The Nit1 promoter allows inducible and reversible silencing of centrin in Chlamydomonas reinhardtii. Eukariotic Cell 2005; 4(11):1959-1962.
64. Poulsen N, Kroger N. A new molecular tool for transgenic diatoms: Control of mRNA and protein biosynthesis by an inducible promoter-terminator cassette. FEBS J 2005; 272(13):3413-3423.
65. Hall LM, Taylor KB, Jones DD. Expression of a foreign gene in Chlamydomonas reinhardtii. Gene 1993; 124(1):75-81.
66. Tang DK, Qiao SY, Wu M. Insertion mutagenesis of Chlamydomonas reinhardtii by electroporation and heterologous DNA. Biochem Molec Biol Int 1995; 36(5):1025-1035.
67. Sizova IA, Lapina TV, Frolova ON et al. Stable nuclear transformation of Chlamydomonas reinhardtii with a Streptomyces rimosus gene as the selectable marker. Gene 1996; 181:13-18.
68. Haring MA, Beck CF. A promoter trap for Chlamydomonas reinhardtii: Development of a gene cloning method using 5' RACE-based probes. Plant J 1997; 11(6):1341-1348.
69. Sizova I, Fuhrmann M, Hegemann P. A Streptomyces rimosus aphVIII gene coding for a new type phosphotransferase provides stable antibiotic resistance to Chlamydomonas reinhardtii. Gene 2001; 277:221-229.
70. Kovar JL, Zhang J, Funke RP et al. Molecular analysis of the acetolactate synthase gene of Chlamydomonas reinhardtii and development of a genetically engineered gene as a dominant selectable marker for genetic transformation. Plant J 2002; 29:109-117.
71. Fischer N, Rochaix JD. The flanking regions of PsaD drive efficient gene expression in the nucleus of the green alga Chlamydomonas reinhardtii. Mol Genet Genomics 2001; 265:888-894.
72. Crutchfield ALM, Diller KR, Brand JJ. Cryopreservation of Chlamydomonas reinhardtii (Chlorophyta). Eur J Phycol 1999; 34:43-52.
73. Baulcombe D. RNA silencing in plants. Nature 2004; 431:356-363.
74. Cerutti H, Johnson AM, Gillham NW et al. A eubacterial gene conferring spectinomycin resistance on Chlamydomonas reinhardtii: Integration into the nuclear genome and gene expression. Genetics 1997; 145:97-110.
75. Cerutti H. RNA interference: Travelling in the cell and gaining function. Trends Genet 2003; 19:39-46.
76. Wu-Scharf D, Jeong B, Zhang C et al. Transgene and transposon silencing in Chlamydomonas by a DEAH-box RNA helicase. Science 2000; 290:1159-1162.
77. Lechtreck KF, Rostmann J, Grunow A. Analysis of Chlamydomonas SF-assemblin by GFP tagging and expression of antisense constructs. J Cell Sci 2002; 115:1511-1522.
78. Jeong B, Wu-Scharf W, Zhang C et al. Suppressors of transcriptional transgenic silencing in Chlamydomonas are sensitive to DNA-damaging agents and reactivate transposable elements. Proc Natl Acad Sci USA 2002; 99:1076-1081.
79. Rohr J, Sarkar N, Balenger S et al. Tandem inverted repeat system for selection of effective transgenic RNAi strains in Chlamydomonas. Plant J 2004; 40:611-621.
80. Fischle W, Wang Y, Allis CD. Histone and chromatin cross-talk. Curr Opin Cell Biol 2003; 15:172-183.
81. Loidl P. A plant dialect of the histone language. Trends Plant Sci 2004; 9:84-90.
82. Van Dijk K, Marley KE, Jeong B et al. Monomethyl histone H3 lysine 4 as an epigenetic mark for silenced euchromatin in Chlamydomonas. Plant Cell 2005; 17.
83. Lumbreras V, Stevens DR, Purton S. Efficient foreign gene expression in Chlamydomonas reinhardtii mediated by an endogenous intron. Plant J 1998; 14:441-447.
84. Zaslavskaia LA, Lippmeier JC, Shih C et al. Trophic conversion of an obligate photoautotrophic organism through metabolic engineering. Science 2001; 292:2073-2075.
85. Mayfield SP, Kindle KL. Stable nuclear transformation of Chlamydomonas reinhardtii by using a C. reinhardtii gene as the selectable marker. Proc Natl Acad Sci USA 1990; 87(6):2087-2091.

CHAPTER 2

Transformation of Cyanobacteria

Agustín Vioque*

Abstract

Cyanobacteria are a diverse and successful group of bacteria defined by their ability to carry out oxygenic photosynthesis. They occupy diverse ecological niches and are important primary producers in the oceans. Cyanobacteria are amenable to genetic manipulation. Some strains are naturally transformable. Many others have been transformed in the lab by conjugation or electroporation. The ability to transform cyanobacteria has been determinant in the development of the molecular biology of these organisms and has been the basis of many of their biotechnological applications. Cyanobacteria are the source of natural products and toxins of potential use and can be engineered to synthesize substances of biotechnological interest. Their high protein and vitamin content makes them useful as a dietary supplement. Because of their ability to occupy diverse ecological niches, they can be used to deliver to the medium substances of interest or as biosensors.

Introduction

The Cyanobacteria are a monophyletic group of organisms, representing one of the more ancient evolutionary lineages among Bacteria. They are all characterized by the ability to carry out oxygenic photosynthesis with two photosystems, similarly to plants. In fact chloroplasts are evolutionarily related to cyanobacteria as they have arisen by an ancient endosymbiotic event, in which a heterotrophic eukaryote engulfed a cyanobacterium-like organism. All present day chloroplasts derive from a single endosymbiotic event that happened around 1.5-1.2 billions years ago.[1] However, evolution of chloroplast containing organisms is complex, and many algal groups, such as diatoms, brown algae or Euglenophytes have arisen by secondary endosymbiosis, in which a chloroplast-less eukaryote engulfed a chloroplast containing eukaryote (see Chapter 3 for more details).[2] It is believed that cyanobacteria are responsible in great part for the increase in oxygen content in the atmosphere of Earth that happened around 2.2 billions years ago, and that they were a necessary requirement for the evolution of complex life forms with aerobic metabolism. Cyanobacteria, thanks to their photoautotrophic mode of growth, can live in poor environments with only light, air, water and some salts. Many strains are able to fix atmospheric dinitrogen into ammonium and therefore they don't need the presence of combined nitrogen in the medium, which makes their culture inexpensive.

Cyanobacteria have a long history of use by man. *Spirulina* is used as a source of protein and vitamins[3,4] because it has a high protein content. *Spirulina platensis* in Africa and *Spirulina maxima* in America have been used as food since antiquity.[3] Other cyanobacteria are used as food in India, China, and the Philippines. Daily eating of *Spirulina* by the Kamemba tribe in Chad is epidemiologically linked to reduced rates of HIV infection.[5]

*Agustín Vioque—Instituto de Bioquímica Vegetal y Fotosíntesis, Universidad de Sevilla-CSIC, Américo Vespucio 49, E-41092 Sevilla, Spain. Email: vioque@us.es

Transgenic Microalgae as Green Cell Factories, edited by Rosa León, Aurora Galván and Emilio Fernández.

Cyanobacteria produce a large variety of secondary metabolites of potential use.[6] Many of these metabolites have complex and unique structures and stereospecificity, which makes their chemical synthesis expensive or impossible. Therefore the cyanobacterium itself must be used as a "cell factory".[7] The pathway for the synthesis of some of these metabolites has been elucidated, and the corresponding genes isolated. Therefore, it is potentially possible to improve their production through the use of transgenic cyanobacteria.

The molecular genetics tools developed for cyanobacteria have been used for the study of many aspects of their biology. A large amount of knowledge on photosynthesis has arisen from studies in cyanobacteria. Detailed excellent reviews on molecular genetics tools for cyanobacteria have been published.[8,9] Here I will summarize briefly the main techniques and describe several lines of research on the applications of cyanobacteria, and specifically of transgenic cyanobacteria.

This review does not intend to be exhaustive, but rather to present selected examples that illustrate on the potential of transgenic cyanobacteria in agriculture, pollution control, and human health.

There are now more than 15 cyanobacteria genomes sequenced, and many more in progress (http://www.kazusa.or.jp/cyano/; http://img.jgi.doe.gov/pub/main.cgi). The availability of genome sequences, which facilitates transcriptomic, proteomic and metabolomic studies is a powerful tool for the development of cyanobacteria as biotechnological tools.

Transformation of Cyanobacteria

The generation of a transgenic organism requires first a procedure for the introduction of the foreign DNA into the cell, and second, the stable maintenance and expression of the introduced DNA. Three procedures are available for the introduction of foreign DNA into cyanobacteria: natural transformation, electroporation, and conjugation.

Natural transformation of cyanobacteria was described more than 30 years ago.[10] However, natural transformation is restricted to a small number of unicellular strains. Most transformable strains are naturally competent. The mechanism of transformation is poorly understood, and it is not known why transformation is possible only with some strains. The reason could be in part the presence of extracellular nucleases that have been found in heterocyst-forming, filamentous strains.[11] The efficiency of DNA uptake in some strains like *Synechococcus* sp. PCC7942 and *Synechocystis* sp. PCC6803 is very high. This high efficiency of transformation has allowed the easy generation of mutants by gene disruption via insertional mutagenesis,[12] and cloning by complementation with genomic libraries.[13,14]

Electroporation has been used for a number of strains, either unicellular or filamentous, but this procedure might be mutagenic.[15,16]

Conjugation has been particularly useful because it is the more general procedure for transforming cyanobacteria. The procedure for conjugational transfer of DNA to cyanobacteria was initially described by Wolk and coworkers[17] but has been extensively improved.[18,19] The DNA to be introduced in the cyanobacterium is cloned in a vector ("cargo" plasmid) that carries a *bom* (*oriT*) site. The mobilization (*mob*) function is provided in trans by a helper plasmid. In addition a conjugative plasmid that carries the transfer functions is required. Usually a triparental mating procedure is used in which an *E.coli* strain carries the vector and the helper plasmids and another strain carries the conjugative plasmid (Fig. 1). Both strains are mixed together with the cyanobacterium, and conjugal transfer of the vector DNA to the cyanobacterium occurs with great efficiency.

A major problem is the presence of restriction systems in the cyanobacterium to be transformed. If the inserted DNA contains the recognition sequence of one of those restriction systems, conjugation efficiency will be dramatically reduced. To alleviate this problem, the helper plasmid can code for methylases that protect against restriction by *Ava*I, *Ava*II, and *Ava*III.[19] Isoschizomers of these enzymes are present in many cyanobacteria.

The foreign DNA inserted in the cyanobacterium will persist in a stable way only if it integrates in the chromosome or in endogenous plasmid by recombination, or if it is capable of

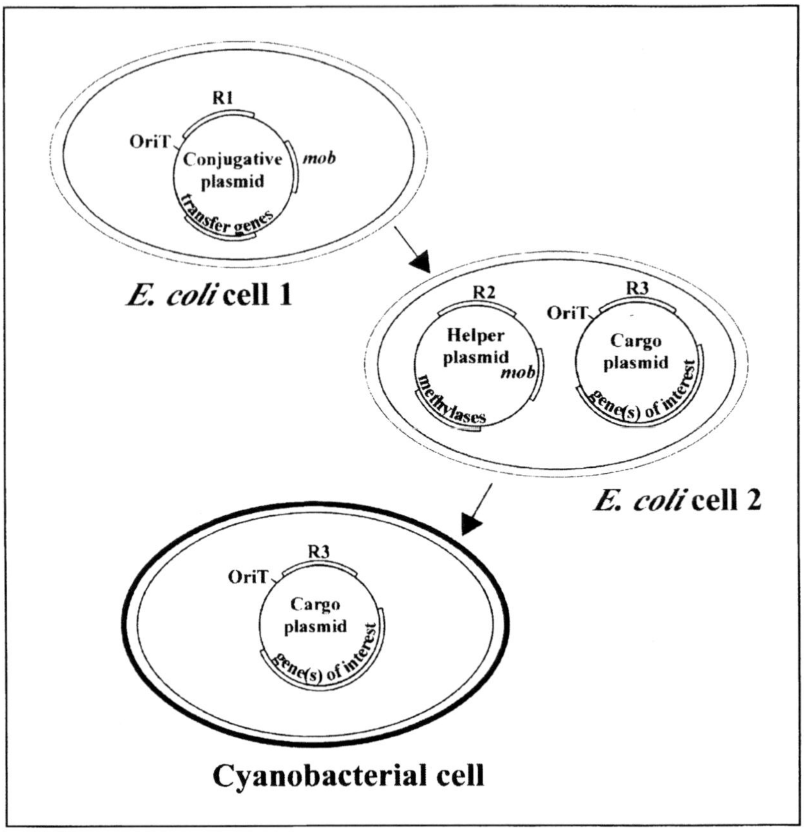

Figure 1. Triparental mating procedure of conjugative DNA transfer to cyanobacterial cells. The two *E. coli* strains and the cyanobacterial cells to be transformed are mixed together. The conjugative plasmid transfers to the *E. coli* cells that carry the cargo and helper plasmids and can mobilize the cargo plasmid that has been cleaved at oriT by the product of the *mob* gene in the helper plasmid. R1, R2, and R3 indicate different antibiotic resistance genes used for selection and maintenance of the plasmids. The cargo plasmid can carry a cyanobacterial replicon or can integrate in the chromosome by recombination.

autonomous replication. DNA can integrate stably in the chromosome through homologous recombination (Fig. 2). Depending on the procedure used, the incoming DNA is circular or linear. Plasmids introduced by conjugation are circular and a single recombination allows integration of the selective marker.[20] Procedures have been developed to select for less frequent double recombinations events if desired.[20,21] When the DNA is introduced directly in those strains that are naturally transformable, double recombinants are mainly selected. This is thought to be due to the cut of the circular DNA during the entry into the cell.[22] Alternatively a vector that can replicate in the cyanobacterium can be used. A whole panoply of vectors, expression platforms, resistance markers, etc have been developed for either insertion in the chromosome or autonomous replication.[8,9,23]

Applications

Some cyanobacteria produce toxins that can make water unpotable. But some of these toxins have shown potential medical application. A large screening of cyanobacterial strains has

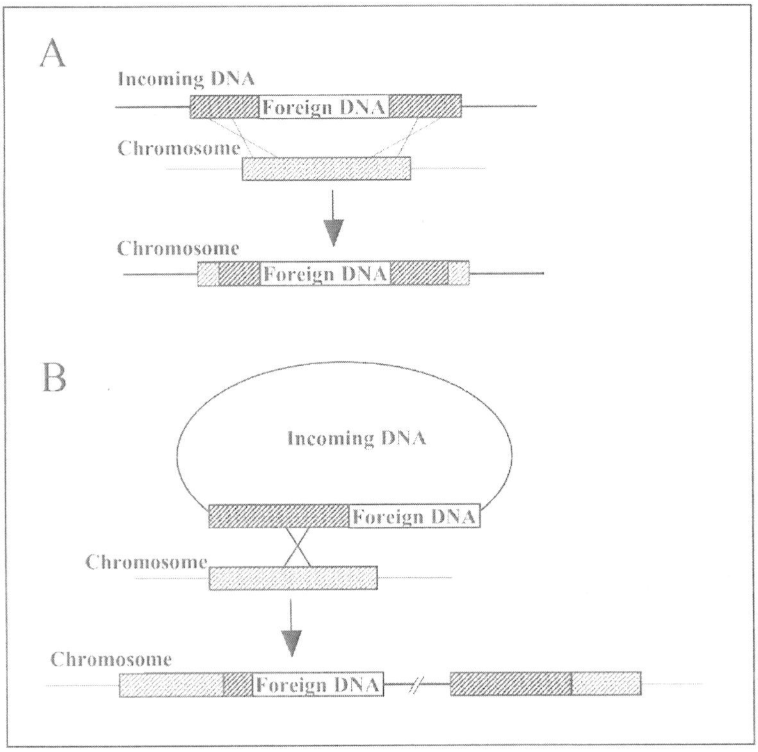

Figure 2. Integration of foreign DNA into cyanobacterial chromosome. Foreign DNA integrates by homologous recombination. A) Double recombination between linear or circular DNA and the chromosome generates the replacement of a wild-type gene by a copy interrupted with the foreign DNA. B) Single recombination between a circular DNA and the chromosome generates a partial diploid. The foreign DNA contains the gene of interest with a promoter for appropriate expression and an antibiotic resistance gene for selection.

detected many compounds with cytotoxic activity that have potential as antineoplasic agents.[6,24] Of special interest is cyanovirin-N, a cyanobacterial lectin discovered in 1997,[25] that is produced by *Nostoc ellipsosporum*. Cyanovirin-N is an 11 kDa peptide with significant antiviral activity against HIV and other enveloped viruses such as Ebola. The mode of action of cyanovirin-N seems to occur through specific interaction with N-linked mannose oligosaccharides on the glycoproteins of the viral surface. This interaction inhibits virus entry. Recombinant cyanovirin-N has been expressed and purified in *E. coli.*[26]

Accumulation of polyhydroxybutirate (PHB) as an intracellular storage compounds has been found in many cyanobacteria. PHB is a biodegradable thermoplastic polyester that could be an interesting substitute of commonly used plastics. Furthermore, cyanobacterial photobiosynthesis of PHB can be used to recycle atmospheric CO_2.[27] Culture conditions that improve PHB accumulation have been described.[28] PHB production has been improved in a *Synechococcus* strain by transposon mutagenesis. A mutant was isolated with increased acetil-CoA, leading to increased PHB accumulation.[29] The genes responsible for PHB synthesis, which include in addition of PHB synthase, an specific β-ketothiolase and an acetoacetyl-CoA reductase, have been identified in *Synechocystis* sp. PCC 6803.[30,31]

Eicosapentaenoic acid (EPA, C20:5) is an ω-3 highly unsaturated fatty acid. This type of fatty acids are receiving attention as useful dietary components because they can prevent several human diseases.[32] The main source of dietary EPA at present is marine fish. Marine cyanobacteria do not synthesize EPA. An EPA synthesis gene cluster from the marine bacterium *Shewanella* sp. SCRC-2738 has been introduced in a marine *Synechococcus* sp. strain by conjugation[33] of a large plasmid (47 kb). The plasmid was not stable due to its large size and the amount of EPA that could be produced was low. The procedure was improved by reducing the size of the plasmid so that it carries only the essential genes required for EPA biosynthesis.[34] Under optimal conditions, EPA represented up to 7.5% of the total fatty acid content of the transconjugant strain.[34]

Cyanophycin is a polypeptide with nonribosomal synthesis that is composed by a polyaspartate backbone and arginine residues linked to the β-carboxyl group of each aspartic acid by their α-amino groups.[35] It has about equimolar amounts of aspartate and arginine. It accumulates as insoluble inclusions in the cytoplasm and serves as an storage compound for carbon, nitrogen, and energy.[36] The biotechnological interest of cyanophycin resides in the fact that it can be chemically converted to poly(aspartic acid) that is a biodegradable substitute for nonbiodegradable polymers.[37] Cyanophycin synthase was purified from *Anabaena variabilis* ATCC29413, and the corresponding gene identified in *Synechocystis* 6803[38] and *Synechocystis* sp. PCC 6308.[39] Its expression in *E.coli* allows cyanophycin synthesis, indicating that it is the only gene required for cyanophycin production. Cyanophycin was thought to be produced only in cyanobacteria, but recently cyanophycin synthase genes have been found in other bacteria.[40] The cyanophycin synthase from the termophilic cyanobacterium *Synechococcus* sp. strain MA19 is of biotechnological interest because of its thermoestability and its relaxed substrate specificity, which allows the synthesis of polymers with different composition.[41]

Carotenoids are natural pigments that are present in all photosynthetic organisms as accessory pigments in light harvesting and in protecting against photo-oxidative damage. In higher plants they provide pigmentation to fruits and flowers and in some animals they contribute to the colors of skin and feathers. They are used in the food industry as colorants and it has been shown that they are good antioxidants that provide protection against free radicals. Of great economical interest is the ketocarotenoid astaxanthin (3,3'-dihydroxy-β, β-carotene-4,4'-dione). The gene responsible of the synthesis of astaxanthin from β-carotene (β-C-4-oxygenase, *crtO*) has been cloned from several organisms that synthesize astaxanthin. The expression of *crtO* from the green alga *Haematococcus pluvialis* in the cyanobacterium *Synechococcus* sp. PCC 7942 resulted in the biosynthesis of astaxanthin in this organisms that is otherwise unable to synthesize it.[42] This experiment showed proof of principle that it is possible to manipulate organisms to accumulate astaxanthin, and has been used to develop transgenic plants which accumulate astaxanthin in edible parts.[43]

The use of cyanobacteria in bioremediation procedures is attractive because they have been found in oil contaminated waters and soil, where they associate with oil-degrading bacteria.[44] An advantage of cyanobacteria over heterotrophic microorganisms is that their growth is not reduced when the pollutant concentration decreases. Hydrocarbon degradation by cyanobacteria has been controversial. In some instances only intracellular hydrocarbon accumulation could be demonstrated. All the hydrocarbon degradation was ascribed to noncyanobacterial associated bacteria.[45] However in other studies it was shown that several strains could degrade hydrocarbons.[46]

Cyanobacteria can degrade halogenated compounds. Lindane (γ-hexachlorocyclohexane) is degraded by *Anabaena* sp. PCC 7120 and *Nostoc ellipsosporum*.[47] Transgenic *Anabaena* carrying the *linA* gene from *Pseudomonas paucimobilis*, which controls the first step of lindane degradation has an increased rate of lindane degradation. Lindane degradation is improved in the presence of nitrate[47] and requires the function of the *nir* operon.[48] 4-chlorobenzoate and 4-iodobenzoate can be degraded by transgenic *Anabaena* sp. PCC 7120 that expresses 4-chlorobenzoate-4-hydroxylase,[47] a dehalogenase from *Arthrobacter globiformis*.[49] *Anabaena* sp. PCC 7120 and *Anabaena flos-aquae* transformed endosulfan, an organochlorine insecticide.[50] Organophosphorus pesticides can also be transformed by cyanobacteria. Some early reports indicated that some strains of cyanobacteria could grow in media supplemented with methyl parathion and other organophosphorus pesticides, but there was no proof of their transformation. Recently, reductive transformation of methyl parathion by *Anabaena* sp. PCC 7120 has been demonstrated under aerobic, photosynthetic conditions.[51] Transformation occurs by reduction of the nitro group to an amino group. The process requires light and is not affected by mutations in the genes involved in nitrate reduction. *Anabaena* strains that have inactivated the *nirA* (nitrite reductase), *nrtC* (nitrate transport), or *narB* (nitrate reductase) genes can transform methyl parathion at a rate similar to wild type,[51] indicating that the assimilatory nitrate reduction system is not involved in methyl parathion transformation, in contrast with the degradation of lindane, that requires a functional *nir* operon.[48]

The accumulation of toxic heavy metals such as copper, cadmium, zinc, lead, mercury or arsenic in soil or water is a serious pollution problem. Efforts have been made for the use of plants or microorganisms that can detoxify the contaminated soil or water through bioaccumulation[52,53] so that expensive physical or chemical remediation procedures can be replaced. Transgenic cyanobacteria expressing metallothionein from yeast[54] or from mouse[55] have been obtained in with the hope that they could be used as bioremediation agents. However their use will require further development and better basic knowledge on metal ion transport and metabolism in cyanobacteria.

Mosquitoes are vectors of many infectious diseases, such as malaria and yellow fever that have important impact on human health. The main procedure to fight these diseases has been the reduction of populations of the transmitting vector through the heavy use of pesticides. The Gram-positive spore-forming bacterium *Bacillus thuringiensis* ssp. *israelensis* (Bti) has been used in the biological control of mosquito populations. Bti produces during sporulation a proteinaceous crystal composed of several insecticidal proteins.[56] The crystal dissolves in the mid-gut of ingesting mosquito larvae and is cleaved into the toxic polypeptides. The primary action of Bti toxins is to lyse midgut epithelial cells in the target insect by forming lytic pores in the apical microvillar membranes.[57] The mosquito larvicidal properties of Bti are due to six proteins in the crystals that act synergistically. The genes that code for these proteins are carried on a plasmid.[58] Application of Bti for mosquito control has found only limited success due to the short residual activity of current preparations under field conditions.[59] A major problem is inactivation by sunlight in the field.[60] Cyanobacteria live in the same niches that mosquito larvae, and are eaten by then.[61] Many efforts have been dedicated to generate transgenic cyanobacteria that express the Bti toxins.[62,63] Due to the synergistic interactions between the toxins, high level of toxicity was observed only when a combination of several proteins was expressed in *Anabaena* sp. PCC 7120.[64] The best level of toxicity has been obtained recently[65] by the coexpression in *Anabaena* sp. PCC 7120 of genes encoding the mosquito larvicidal toxins Cry4Aa, Cry11Aa, Cyt1Aa and the regulatory protein P20 under the control of the strong constitutive promoters P_{psbA} and P_{A1}. Transgenic *Anabaena* sp. PCC 7120 has been shown to be superior in the field compared to other procedures of delivering the Bti toxins.[66,67] One of the reasons is that expression in cyanobacteria protects the toxins against fast inactivation by sunlight in the field.[68]

In a general sense biosensors are devices based on biological material for the quantitative detection of substances. On a more specific sense, biosensors are whole organisms that can give a dose dependent detectable response to the presence of a chemical in the medium. Biosensors are used to detect the presence of pollutants, pesticides or other chemicals of interest in the medium. Of course similar detection can be achieved by chemical analysis, but biosensors allow the estimation of the concentrations of biologically available, which is quite often the relevant parameter, rather than total concentrationand no sophisticated instrumentation is required. Also biosensors can be designed to detect whole classes of substances, or alternatively they can be highly specific thanks to the specificity on the interactions used (enzyme-substrate, transcription factor-promoter, nucleic acid complementarity, etc). There are two possible designs of biosensors (Fig. 3). In one design, the substance(s) to be detected inhibits some aspect

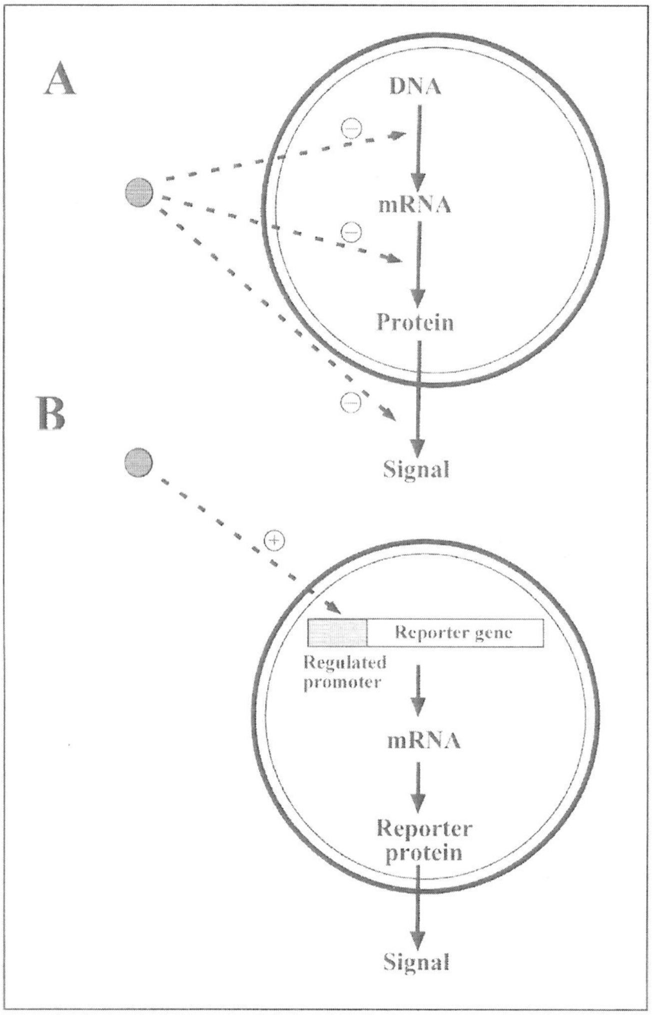

Figure 3. Cyanobacteria as biosensors. A) The concentration of a chemical is estimated from the inhibition of metabolic activity at any level of cellular metabolism, resulting in a quantifiable reduction of a detectable signal. B) The reporter gene is fused to a promoter that is activated by the chemical. The grey circle represents the chemical to be detected.

of cell metabolism (Fig. 3A). This design has been called "light on".[69] The metabolic inhibition is coupled to an easily detectable response. Another possible design is through a reporter gene under the control of a promoter that is activated by the presence of the substance to be detected (Fig. 3B). This design has been named "light off".[69] The light on design has been used to develop transgenic cyanobacteria that detect herbicides and other pollutants that inhibit metabolism by the use of luciferase as a reporter. Bioluminescence produced by luciferase activity is directly dependent on the intracellular ATP concentration, which is sensitive to photosynthetic activity. Any reduction in photosynthesis due to the presence of herbicides or other toxic substances results in a reduction in ATP level, which in turns reduces bioluminescence due to luciferase. For instance the luciferase *luc* gene from the firefly *Photinus pyralis* was introduced in

Synechocystis sp. PCC6803[70] under the control of the *tac* promoter. Bioluminescence of this *Synechocystis* strain was very sensitive to a broad range of herbicides and other pollutants like heavy metals and volatile organic pollutants. Several cyanobacterial biosensors based on reporter genes under the control of a promoter activated by the factor to be detected (Fig. 3B) have been developed. These biosensors have been used for the detection of bioavailability of important nutrients such as phosphorus, iron or nitrate. A nitrate biosensor was made to monitor nitrate bioavailability in aquatic ecosystems using a *Synechocystis* strain carrying in the chromosome an insertion of the luciferase operon from *Vibrio harveyi* (*luxAB*) under the control of the *nblA* promoter.[71] Cyanobacteria have developed adaptations to cope with environmental changes.[72] In cells deprived of essential nutrients the phycobilisomes are rapidly degraded by an ordered proteolytic process.[73] NblA protein is essential for phycobilisome degradation,[74] and its expression is up-regulated under nitrate deprivation. The transgenic *Synechocystis* cells carrying the *PnblA::luxAB* fusion were immobilized in microtiter plates and showed a dose-dependent response in the range 4-100 μM nitrate.[71] A similar approach was used in which the *luxAB* gene was fused to the *glnA* (glutamine synthetase) promoter and the construction was introduced in a neutral site in the genome of *Synechococcus* sp. PCC 7942.[75] The *glnA* promoter is responsive to the available nitrogen in the medium.[76] It was shown that the bioluminescence is inversely proportional to the N concentration in the range 1 μM to 1 mM for ammonium, nitrate or nitrite and 10- to 50- fold higher for the organic N compounds glutamine and urea. Phosphorus bioavailability was measured with the promoter of *phoA* (alkaline phosphatase) from *Synechococcus* sp. PCC 7942 fused to *luxAB*.[77] The expression of *phoA* is repressed by available phosphate in the medium. Transgenic *Synechococcus* carrying the *PphoA::luxAB* fusion integrated in the chromosome emitted light when inorganic P concentrations fell below 2.3 μM. Light emission decreased with increased P concentration in the medium. In a natural freshwater environment luminescence correlated with dissolved phosphate concentration. As a final recent example, a *Synechococcus* sp. PCC 7942 biosensor has also been made for the detection of bioavailable iron by the use of the Fe-responsive *isiAB* promoter fused to *luxAB*.[78] In order to provide the aldehyde substrate of luciferase in vivo, the *luxCDE* genes, driven by the *Synechococcus psbAI* promoter were integrated at a second neutral site. Although the lineal response to Fe concentration was limited to a short range around 10^{-21} M, it was useful to estimate bioavailability of Fe in natural water samples.[79]

Acknowledgements

Grants from Ministerio de Educación y Ciencia (BFU2004-00076/BMC) and Junta de Andalucía (CVI-215) support research in the author lab.

References

1. Rodriguez-Ezpeleta N, Brinkmann H, Burey SC et al. Monophyly of primary photosynthetic eukaryotes: Green plants, red algae, and glaucophytes. Curr Biol 2005; 15:1325-1330.
2. Bhattacharya D, Yoon HS, Hackett JD. Photosynthetic eukaryotes unite: Endosymbiosis connects the dots. Bioessays 2004; 26:50-60.
3. Ciferri O. Spirulina, the edible microorganism. Microbiol Rev 1983; 47:551-578.
4. Kay RA. Microalgae as food and supplement. Crit Rev Food Sci Nutr 1991; 30:555-573.
5. Teas J, Hebert JR, Fitton JH et al. Algae — A poor man's HAART? Med Hypotheses 2004; 62:507-510.
6. Burja AM, Banaigs B, Abou-Mansour E et al. Marine cyanobacteria-a prolific source of natural products. Tetrahedron 2001; 57:9347-9377.
7. Burja AM, Dhamwichukorn S, Wright PC. Cyanobacterial postgenomic research and systems biology. Trends Biotechnol 2003; 21:504-511.
8. Thiel T. Genetic analysis of cyanobacteria. In: Bryant DA, ed. The Molecular Biology of Cyanobacteria. Vol 1. Dordrecht: Kluwer Academic Publishers, 1994:581-611.
9. Elhai J. Genetic techniques appropriate for the biotechnological exploitation of cyanobacteria. J Appl Phycol 1994; 6:177-186.
10. Shestakov SV, Khyen NT. Evidence for genetic transformation in blue-green alga Anacystis nidulans. Mol Gen Genet 1970; 107:372-375.

11. Wolk CP, Kraus J. Two approaches to obtaining low, extracellular deoxyribonuclease activity in cultures of heterocyst-forming cyanobacteria. Arch Microbiol 1982; 131:302-307.
12. Chauvat F, Rouet P, Bottin H et al. Mutagenesis by random cloning of an Escherichia coli kanamycin resistance gene into the genome of the cyanobacterium Synechocystis PCC 6803: Selection of mutants defective in photosynthesis. Mol Gen Genet 1989; 216:51-59.
13. Dzelzkalns VA, Bogorad L. Molecular analysis of a mutant defective in photosynthetic oxygen evolution and isolation of a complementing clone by a novel screening procedure. EMBO J 1988; 7:333-338.
14. Martinez-Ferez IM, Vioque A. Nucleotide sequence of the phytoene desaturase gene from Synechocystis sp. PCC 6803 and characterization of a new mutation which confers resistance to the herbicide norflurazon. Plant Mol Biol 1992; 18:981-983.
15. Muhlenhoff U, Chauvat F. Gene transfer and manipulation in the thermophilic cyanobacterium Synechococcus elongatus. Mol Gen Genet 1996; 252:93-100.
16. Bruns BU, Briggs WR, Grossman AR. Molecular characterization of phycobilisome regulatory mutants of Fremyella diplosiphon. J Bacteriol 1989; 171:901-908.
17. Wolk CP, Vonshak A, Kehoe P et al. Construction of shuttle vectors capable of conjugative transfer from Escherichia coli to nitrogen-fixing filamentous cyanobacteria. Proc Natl Acad Sci USA 1984; 81:1561-1565.
18. Thiel T, Wolk CP. Conjugal transfer of plasmids to cyanobacteria. Methods Enzymol 1987; 153:232-243.
19. Elhai J, Wolk CP. Conjugal transfer of DNA to cyanobacteria. Methods Enzymol 1988; 167:747-754.
20. Cai YP, Wolk CP. Use of a conditionally lethal gene in Anabaena sp. strain PCC 7120 to select for double recombinants and to entrap insertion sequences. J Bacteriol 1990; 172:3138-3145.
21. Black TA, Cai Y, Wolk CP. Spatial expression and autoregulation of hetR, a gene involved in the control of heterocyst development in Anabaena. Mol Microbiol 1993; 9:77-84.
22. Porter RD. Transformation in cyanobacteria. CRC Crit Rev Microbiol 1987; 13:111-132.
23. Koksharova OA, Wolk CP. Genetic tools for cyanobacteria. Appl Microbiol Biotechnol 2002; 58:123-137.
24. Patterson MLG, Baldwin CL, Bolis CM et al. Antineoplastic activity of cultured blue- green algae (Cyanophyta). J Phycol 1991; 27:530-536.
25. Boyd MR, Gustafson KR, McMahon JB et al. Discovery of cyanovirin-N, a novel human immunodeficiency virus-inactivating protein that binds viral surface envelope glycoprotein gp120: Potential applications to microbicide development. Antimicrob Agents Chemother 1997; 41:1521-1530.
26. Colleluori DM, Tien D, Kang F et al. Expression, purification, and characterization of recombinant cyanovirin-N for vaginal anti-HIV microbicide development. Protein Expr Purif 2005; 39:229-236.
27. Asada Y, Miyake M, Miyake J et al. Photosynthetic accumulation of poly-(hydroxybutyrate) by cyanobacteria—the metabolism and potential for CO2 recycling. Int J Biol Macromol 1999; 25:37-42.
28. Sharma L, Mallick N. Enhancement of poly-beta-hydroxybutyrate accumulation in Nostoc muscorum under mixotrophy, chemoheterotrophy and limitations of gas-exchange. Biotechnol Lett 2005; 27:59-62.
29. Miyake M, Takase K, Narato M et al. Polyhydroxybutyrate production from carbon dioxide by cyanobacteria. Appl Biochem Biotechnol 2000; 84-86:991-1002.
30. Taroncher-Oldenburg G, Nishina K, Stephanopoulos G. Identification and analysis of the polyhydroxyalkanoate-specific beta-ketothiolase and acetoacetyl coenzyme A reductase genes in the cyanobacterium Synechocystis sp. strain PCC6803. Appl Environ Microbiol 2000; 66:4440-4448.
31. Hein S, Tran H, Steinbuchel A. Synechocystis sp. PCC6803 possesses a two-component polyhydroxyalkanoic acid synthase similar to that of anoxygenic purple sulfur bacteria. Arch Microbiol 1998; 170:162-170.
32. Siddiqui RA, Shaikh SR, Sech LA et al. Omega 3-fatty acids: Health benefits and cellular mechanisms of action. Mini Rev Med Chem 2004; 4:859-871.
33. Takeyama H, Takeda D, Yazawa K et al. Expression of the eicosapentaenoic acid synthesis gene cluster from Shewanella sp. in a transgenic marine cyanobacterium, Synechococcus sp. Microbiology 1997; 143:2725-2731.
34. Yu R, Yamada A, Watanabe K et al. Production of eicosapentaenoic acid by a recombinant marine cyanobacterium, Synechococcus sp. Lipids 2000; 35:1061-1064.
35. Simon RD, Weathers P. Determination of the structure of the novel polypeptide containing aspartic acid and arginine which is found in Cyanobacteria. Biochim Biophys Acta 1976; 420:165-176.

36. Li H, Sherman DM, Bao S et al. Pattern of cyanophycin accumulation in nitrogen-fixing and nonnitrogen-fixing cyanobacteria. Arch Microbiol 2001; 176:9-18.
37. Schwamborn M. Chemical synthesis of polyaspartates: A biodegradable alternative to currently used polycarboxylate homo- and copolymers. Polym Degrad Stabil 1998; 59:39-45.
38. Ziegler K, Diener A, Herpin C et al. Molecular characterization of cyanophycin synthetase, the enzyme catalyzing the biosynthesis of the cyanobacterial reserve material multi-L-arginyl-poly-L-aspartate (cyanophycin). Eur J Biochem 1998; 254:154-159.
39. Aboulmagd E, Oppermann-Sanio FB, Steinbuchel A. Molecular characterization of the cyanophycin synthetase from Synechocystis sp. strain PCC6308. Arch Microbiol 2000; 174:297-306.
40. Krehenbrink M, Oppermann-Sanio FB, Steinbuchel A. Evaluation of noncyanobacterial genome sequences for occurrence of genes encoding proteins homologous to cyanophycin synthetase and cloning of an active cyanophycin synthetase from Acinetobacter sp. strain DSM 587. Arch Microbiol 2002; 177:371-380.
41. Hai T, Oppermann-Sanio FB, Steinbuchel A. Molecular characterization of a thermostable cyanophycin synthetase from the thermophilic cyanobacterium Synechococcus sp. strain MA19 and in vitro synthesis of cyanophycin and related polyamides. Appl Environ Microbiol 2002; 68:93-101.
42. Harker M, Hirschberg J. Biosynthesis of ketocarotenoids in transgenic cyanobacteria expressing the algal gene for beta-C-4-oxygenase, crtO. FEBS Lett 1997; 404:129-134.
43. Mann V, Harker M, Pecker I et al. Metabolic engineering of astaxanthin production in tobacco flowers. Nat Biotechnol 2000; 18:888-892.
44. Sorkhoh N, Al-Hasan R, Radwan S et al. Self-cleaning of the Gulf. Nature 1992; 359:109.
45. Chaillan F, Gugger M, Saliot A et al. Role of cyanobacteria in the biodegradation of crude oil by a tropical cyanobacterial mat. Chemosphere 2005, (in press).
46. Raghukumar C, Vipparty V, David JJ et al. Degradation of crude oil by marine cyanobacteria. Appl Microbiol Biotechnol 2001; 57:433-436.
47. Kuritz T, Wolk CP. Use of filamentous cyanobacteria for biodegradation of organic pollutants. Appl Environ Microbiol 1995; 61:234-238.
48. Kuritz T, Bocanera LV, Rivera NS. Dechlorination of lindane by the cyanobacterium Anabaena sp. strain PCC7120 depends on the function of the nir operon. J Bacteriol 1997; 179:3368-3370.
49. Tsoi TV, Zaitsev GM, Plotnikova EG et al. Cloning and expression of the Arthrobacter globiformis KZT1 fcbA gene encoding dehalogenase (4-chlorobenzoate-4-hydroxylase) in Escherichia coli. FEMS Microbiol Lett 1991; 65:165-169.
50. Lee SE, Kim JS, Kennedy IR et al. Biotransformation of an organochlorine insecticide, endosulfan, by Anabaena species. J Agric Food Chem 2003; 51:1336-1340.
51. Barton JW, Kuritz T, O'Connor LE et al. Reductive transformation of methyl parathion by the cyanobacterium Anabaena sp. strain PCC7120. Appl Microbiol Biotechnol 2004; 65:330-335.
52. Meagher RB. Phytoremediation of toxic elemental and organic pollutants. Curr Opin Plant Biol 2000; 3:153-162.
53. Lloyd JR, Lovley DR, Macaskie LE. Biotechnological application of metal-reducing microorganisms. Adv Appl Microbiol 2003; 53:85-128.
54. McCormick PM, Cannon GC, Heinhorst S. Expression of the copper metallothionein CUPI from Saccharomyces cerevisiae in the cyanobacterium Synechococcus R2-PIM8(smtA). Curr Microbiol 1999; 38:155-162.
55. Shao Q, Shi DJ, Hao FY et al. Cloning and expression of metallothionein mutant alpha-KKS-alpha in Anabaena sp. PCC 7120. Mar Pollut Bull 2002; 45:163-167.
56. de Maagd RA, Bravo A, Berry C et al. Structure, diversity, and evolution of protein toxins from spore-forming entomopathogenic bacteria. Annu Rev Genet 2003; 37:409-433.
57. Schnepf HE, Crickmore N, Van Rie J et al. Bacillus thuringiensis and its pesticidal crystal proteins. Microbiol Mol Biol Rev 1998; 62:775-806.
58. Berry C, O'Neil S, Ben-Dov E et al. Complete sequence and organization of pBtoxis, the toxin-coding plasmid of Bacillus thuringiensis subsp. israelensis. Appl Environ Microbiol 2002; 68:5082-5095.
59. Margalith Y, Ben-Dov E. Biological Control by Bacillus thuringiensis subsp. israelensis. In: Rechcigl JE, Rechcigl NA, eds. Insect Pest Management: Techniques for Environmental Protection. Boca Raton: CRC Press, 2000:243-301.
60. Pusztai M, Fast P, Gringorten L et al. The mechanism of sunlight-mediated inactivation of Bacillus thuringiensis crystals. Biochem J 1991; 273:43-47.
61. Thiery I, Nicolas L, Rippka R et al. Selection of cyanobacteria isolated from mosquito breeding sites as a potential food source for mosquito larvae. Appl Environ Microbiol 1991; 57:1354-1359.
62. Tandeau de Marsac N, de la Torre F, Szulmajster J. Expression of the larvicidal gene of Bacillus sphaericus 1593M in the cyanobacterium Anacystis nidulans R2. Mol Gen Genet 1987; 209:396-398.

63. Murphy RC, Stevens Jr SE. Cloning and expression of the cryIVD gene of Bacillus thuringiensis subsp. israelensis in the cyanobacterium Agmenellum quadruplicatum PR-6 and its resulting larvicidal activity. Appl Environ Microbiol 1992; 58:1650-1655.
64. Wu X, Vennison SJ, Huirong L et al. Mosquito larvicidal activity of transgenic Anabaena strain PCC 7120 expressing combinations of genes from Bacillus thuringiensis subsp. israelensis. Appl Environ Microbiol 1997; 63:4971-4974.
65. Khasdan V, Ben-Dov E, Manasherob R et al. Mosquito larvicidal activity of transgenic Anabaena PCC 7120 expressing toxin genes from Bacillus thuringiensis subsp. israelensis. FEMS Microbiol Lett 2003; 227:189-195.
66. Manasherob R, Otieno-Ayayo ZN, Ben-Dov E et al. Enduring toxicity of transgenic Anabaena PCC 7120 expressing mosquito larvicidal genes from Bacillus thuringiensis ssp. israelensis. Environ Microbiol 2003; 5:997-1001.
67. Lluisma AO, Karmacharya N, Zarka A et al. Suitability of Anabaena PCC7120 expressing mosquitocidal toxin genes from Bacillus thuringiensis subsp. israelensis for biotechnological application. Appl Microbiol Biotechnol 2001; 57:161-166.
68. Manasherob R, Ben-Dov E, Xiaoqiang W et al. Protection from UV-B damage of mosquito larvicidal toxins from Bacillus thuringiensis subsp. israelensis expressed in Anabaena PCC 7120. Curr Microbiol 2002; 45:217-220.
69. Belkin S. Microbial whole-cell sensing systems of environmental pollutants. Curr Opin Microbiol 2003; 6:206-212.
70. Shao CY, Howe CJ, Porter AJ et al. Novel cyanobacterial biosensor for detection of herbicides. Appl Environ Microbiol 2002; 68:5026-5033.
71. Mbeunkui F, Richaud C, Etienne AL et al. Bioavailable nitrate detection in water by an inmobilized luminescent cyanobacterial reporter strain. Appl Microbiol Biotechnol 2002; 60:306-312.
72. Tandeau de Marsac N, Houmard J. Adaptation of cyanobacteria to environmental stimuli: New steps towards molecular mechanisms. FEMS Microbiol Rev 1993; 104:119-190.
73. Grossman AR, Bhaya D, Apt KE et al. Light-harvesting complexes in oxygenic photosynthesis: Diversity, control, and evolution. Annu Rev Genet 1995; 29:231-288.
74. Collier JL, Grossman AR. Chlorosis induced by nutrient deprivation in Synechococcus sp. strain PCC 7942: Not all bleaching is the same. J Bacteriol 1992; 174:4718-4726.
75. Gillor O, Harush A, Hadas O et al. A Synechococcus PglnA:luxAB fusion for estimation of nitrogen bioavailability to freshwater cyanobacteria. Appl Environ Microbiol 2003; 69:1465-1474.
76. Muro-Pastor MI, Reyes JC, Florencio FJ. Ammonium assimilation in cyanobacteria. Photosynth Res 2005; 83:135-150.
77. Gillor O, Hadas O, Post AF et al. Phosphorus bioavailability monitoring by a bioluminiscent cyanobacterial sensor strain. J Phycol 2002; 38:107-115.
78. Durham KA, Porta D, Twiss MR et al. Construction and initial characterization of a luminescent Synechococcus sp. PCC 7942 Fe-dependent bioreporter. FEMS Microbiol Lett 2002; 209:215-221.
79. Porta D, Bullerjahn GS, Durham KA et al. Physiological characterization of a Synechococcus sp. (Cyanophyceae) strain PCC 7942 iron-dependent bioreporter for freshwater environments. J Phycol 2003; 39:64-73.

Chapter 3

Molecular Biology and the Biotechnological Potential of Diatoms

Peter Kroth*

Abstract

Diatoms are unicellular photoautotrophic eukaryotes that play an important role in ecology by fixing large amounts of CO_2 in the oceans. Because they evolved by secondary endocytobiosis—a process of uptake of a eukaryotic alga into another eukaryotic cell—they have a rather unusual cell biology and genetic constitution. Diatoms are also of biotechnological interest since they produce highly unsaturated fatty acids. In addition they are able to form delicately ornate cell walls made of amorphous silica. Understanding and modifying the processes of biomineralization in diatoms might result in new nanotechnological processes. Therefore recent advances in molecular genomics and the development of genetic tools for diatoms might pave the way for biotechnological modification and utilization of diatoms. In this chapter I will briefly characterize these extraordinary organisms, give some insights into the actual advances in molecular biology of diatoms and present some examples for the potential future use of diatoms in algal biotechnology.

Diatom Biology

Diatoms have fascinated botanists since they were first discovered by light microscopy and mistakenly addressed as unicellular animals due to their brownish colour and their capability to move on substrates.[1] Diatoms are single celled, sometimes colonial organisms belonging to the class Bacillariophyceae of the phylum Bacillariophyta. Almost all of them are photoautotrophic and can be found in nearly any aquatic and even in some terrestrial habitats.[2] Although they are a phylogenetically rather young group—they first appeared about 180 million years ago[3]—there is a huge diversity of some estimated 10.000 - 100.000 different species[4] in about 250 genera.[5] This rather broad estimate is due to difficulties in identification and separation of individual diatom species. According to the species concept, organisms mating with each other belong to the same species. Unfortunately our knowledge about diatom sexuality is rather poor, because the induction of reproductive stages in culture frequently failed.[6] Diatom taxonomy thus is based mainly either on the identification of ribosomal sequences[7] or—more classically—on the morphology and the shape of the frustules, the extracellular silica cell walls. These frustules, which are formed by two valves that fit together like Petri dishes, often are highly ornamented and show species-specific structures (Fig. 1). The ability of diatoms to genetically define these structures makes them highly interesting for nanotechnological applications.[8] Because of their siliceous composition frustules are often well preserved in

*Peter Kroth—Fachbereich Biologie, Universität Konstanz, 78457 Konstanz, Germany.
Email: Peter.Kroth@uni-konstanz.de

Transgenic Microalgae as Green Cell Factories, edited by Rosa León, Aurora Galván and Emilio Fernández.

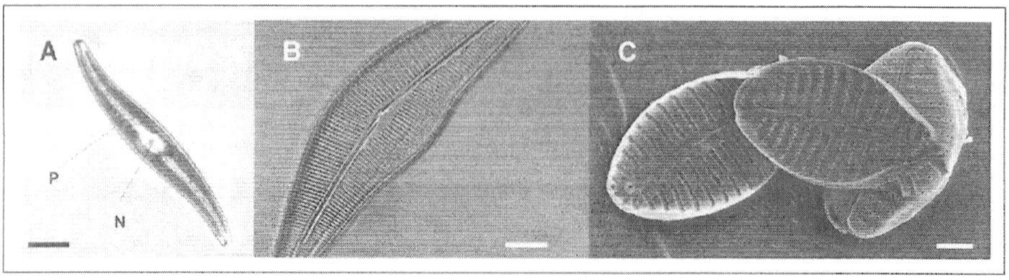

Figure 1. Light and electron microscopical images of diatoms. A) Living cell of *Gyrosigma* sp.: The long dark structures represent the plastids (P), the nucleus (N) is located in the center of the cell, the scale bar represents 16 μm. B) Valve structure of *Cymbella lanceolata* after cleaning with sulphuric acid, the scale bar represents 3 μm (by courtesy of Rahul Bahulikar, Plant Ecophysiology, University of Konstanz). C) Scanning electron micrograph of cleaned valves of *Cymbella microcephala* showing the interior and the surface area. The scale bar represents 1 μm (by courtesy of Kurt Mendgen, Phytopathology, University of Konstanz)

fossil deposits.[9] Strong sedimentation of diatom shells in ancient oceans led to the deposition of siliceous earth or diatomite, a material of high importance for industrial uses.[10]

Diatoms play a significant ecological role. About half of the global annual net primary production in the oceans is due to phytoplankton which is dominated by diatoms.[11] Diatoms thus produce and represent the main input into the marine food web. They are not only strongly involved in fixation of CO_2, but also in cycling of soluble silicates by integrating them into the shells and by releasing parts of them after decomposition at the bottom of the oceans.[12] It is not yet fully understood why diatoms filled aquatic niches so successfully, even though they have the capability to grow at a wide range of light intensities[13] and the apparent ability to perform C4 photosynthesis.[14]

Another peculiar aspect of diatoms is their evolution. The phylogenetically inaccurate term "algae" applies to a variety of organisms showing a broad diversity. Nonetheless, various analyses clearly indicate that the plastids of all the different algae we know today may be traced back to a single endosymbiosis event ("primary endosymbiosis" or more correctly "primary endocytobiosis"), a process in which a cyanobacterium was taken up by a eukaryotic host cell and subsequently transformed into a plastid (Fig. 2).[15] An apparently inevitable consequence of this "domestication" is the transfer of most of the endosymbiont's genes into the nucleus of the host cell. As we know today, several algal groups including the diatoms are supposed to have evolved in an even more complex way by secondary endocytobiosis, a process that successfully happened at least twice during evolution (Fig. 2).[15,16] Here, eukaryotic algae, themselves possessing plastids, were taken up by host cells and were transformed into plastids. Due to redundancy, nearly all of the cytoplasmic structures of the endosymbiotic algae have vanished—usually including their nuclear genome—while the plastids have been preserved; probably because of their ability to perform photosynthesis they must have been highly attractive for the host cell. Diatom plastids are also termed 'complex' plastids, because in comparison to land plant plastids they possess two additional surrounding membranes probably originating from the phagotrophic membrane of the host and the plasma membrane of the endosymbiont.[15] In diatoms the outermost plastid membrane contains ribosomes and is connected to the cytosolic endoplasmic reticulum (ER), the plastids are thus essentially located in a sac of ER (Fig. 2).[17] Due to gene transfer processes during secondary endocytobiosis, diatom nuclei may possess genes of both primary and secondary host cells (including their mitochondria) as well as of the plastid. The recent sequencing project exploring the genome of the diatom *Thalassiosira pseudonana*[18] supported the expected genetic complexity of diatoms indicating that only half of the genes encode proteins similar to red algae, green algae and plants, while about 7% of genes encode"animal"-like proteins that may been derived from the secondary host cell.[18]

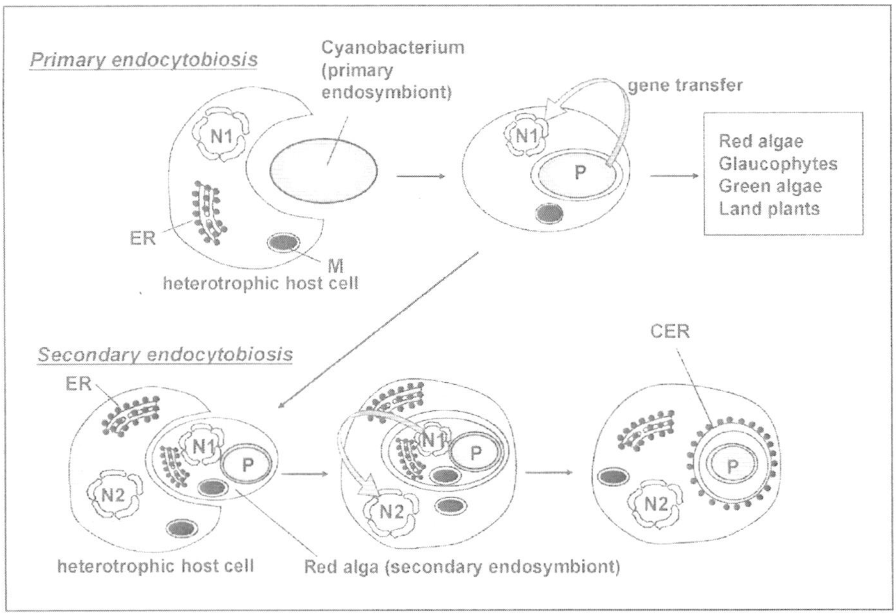

Figure 2. Putative processes involved in primary and secondary endocytobiosis. *Primary endocytobiosis*: A phototrophic cyanobacterium is engulfed by a heterotrophic eukaryotic cell and transformed into a primary plastid surrounded by two envelope membranes. During this process, most of the cyanobacterial genome was transferred to the nucleus (N1) of the host cell (indicated by a bent arrow) or simply got lost. The resulting so-called primary alga may be regarded as the prototype of eukaryotic algae and is represented by modern glaucophytes, red algae and green algae. *Secondary endocytobiosis*: A eukaryotic primary alga containing a chloroplast (P) is taken up by a heterotrophic eukaryotic cell and subsequently transformed into a secondary plastid surrounded by four membranes. Transformation includes massive gene transfer from the algal nucleus (N1) to the secondary host nucleus (N2) (indicated by a bent arrow). Finally, the highly reduced endosymbiont is completely integrated in the host cell as a secondary plastid. In the case of diatoms the outermost plastid membrane contains bound ribosomes ("chloroplast ER", CER).

Genetic Manipulation of Diatoms

Diatoms are almost unique among algae in being diplontic, and sexual reproduction is an obligate stage in the life cycle of most diatom species.[6] Most diatoms, however, often behave asexually in culture, because depending on the species sexual reproduction only occurs under certain environmental conditions or when the cells reach a certain critical size.[19] Genetic proofs for actual crosses of transgenic diatom lineages unfortunately have not yet been published, limiting the genetic methods available.

There are several ways to analyze and optimize organisms that produce substances of biotechnological interest. One possibility is to screen natural populations of the individual organisms for strains that show a higher production within the natural variation. Another approach is random mutagenesis by UV or chemicals, followed by a selection of strains with a higher productivity. While these approaches basically rely on the selection of phenotypes, genetic transformation of diatoms allows a targeted modification of the genotype of the respective organisms. Genetic transformation (or transfection) by definition means the stable incorporation of foreign genetic material into the genome of a cell. Significant progress on gene transfer systems in eukaryotic algae has mainly been achieved in the last 20 years allowing the modification of algae either in order to obtain strains which produce certain compounds of commercial interest or to gain information about cellular, physiological or biochemical mechanisms by switching off, downregulating or overexpressing existing or foreign genes, respectively (see also Chapters 1, 4, and 5).

Although the genome of one diatom (*Thalassiosira pseudonana*) is completely sequenced,[18] the genome of another (*Phaeodactylum tricornutum*) has recently been annotated (Bowler et al, unpublished, http://genome.jgi-psf.org/Phatr1/Phatr1.home.html), and there is also a large set of EST data available,[20] genetic manipulation of diatoms is still in its infancy. For a long time molecular biology of eukaryotic algae was focused on a single organism: the unicellular green alga *Chlamydomonas reinhardtii*. This microalga is still the favorite model system for studying photosynthesis, chloroplast biogenesis, and flagellar function and assembly because of its well-defined genetics.[21] However, in the last ten years basic genetic transformation systems for diatoms have been developed. The genetic information within eukaryotic algae is located in the nucleus, the chloroplasts and the mitochondria. While in *Chlamydomonas* all the three respective genomes meanwhile can be transformed genetically,[22,23] for diatoms there are only reports about nuclear transformation systems.[24-27] Just ten years ago Dunahay et al were the first to report a genetic transformation system for the diatoms *Cyclotella cryptica* and *Navicula saprophila*.[24] They used vectors containing the acetyl-CoA carboxylase from *Cyclotella* to drive expression of the neomycin phosphotransferase gene for selection. Soon thereafter reports appeared describing transformation protocols for *Phaeodactylum tricornutum*[25,26] and *Cylindrotheca fusiformis*[28] (see Table 1). A general purpose transformation vector was constructed for *Phaeodactylum* possessing a resistance cassette and a multiple cloning site for the uptake of reporter genes.[29] In all described protocols diatom cells were transformed by bombardment using a biolistic device, which already had been successfully used for transformation of unicellular microbes, plants and mammalian tissue.[30] For this procedure DNA is precipitated on tungsten particles of defined sizes, and the particles are subsequently shot onto the diatoms which before have been spread in a monolayer on an agar plate. Other procedures that were successfully deployed in *Chlamydomonas* like shaking of the cells with glass beads[31] or silica carbide whiskers[32] and electroporation[33] have not yet been successful in diatoms.

The choice of the right diatom for transformation experiments can be difficult: while researchers interested in basic research might be looking for a model organism representing general aspects, projects focused on the optimization of a strain used for commercial production do not have this choice. Generally, it should be possible to extend biolistic transformation to other diatom species; however there are certain restrictions: First the diatom strain of choice has to be able to grow reasonably well under laboratory conditions. At a division rate of one per day it already takes 2 to 3 weeks for colonies to show up; lower division rates inevitably would prolong screening and might result in decreased efficiencies of the antibiotics used for selection. Next it should be possible to maintain the strain axenically, otherwise it might become overgrown quickly by contaminating organisms after transformation especially if the growth rate of the diatom is poor. As all protocols so far rely on the selection of colonies on solid media, the diatom of choice should be capable of growing on agar plates and ideally should be immobile. The standard selection methods utilize antibiotics, some of which are unstable over a longer period at given culture conditions or in the light. If the diatom grows rather slowly, the induction of growth inhibition or cell death by the selective antibiotic could be weakened, giving false positive colonies. As the gradual size reduction during vegetative multiplication in diatoms can result in an ultimate loss of a clonal culture,[6] those diatom species that undergo only a minimal size reduction should be preferred.

The process of nuclear DNA transformation relies on random integration of modified DNA into the nuclear genome. At first glance this appears to be disadvantageous, because each transformed cell line is different and it cannot be excluded that wild type genes may have been modified at the site of integration. On the other hand with each transformation experiment different transformants are produced which express the gene of interest with different intensities,

Table 1. Diatoms used for genetic manipulation and genome analysis

Diatom	Transformation Protocol Available	Genome Available	Promoter	Expressed Reporter genes	Resistance Cassette	Reference
Cyclotella cryptica	+	-	acc	-	nptII	24
Cylindrotheca fusiformis	+	-	fcp	frue, Hup1	ble	28
				nr		27
Navicula saprophila	+	-	acc	-	nptII	24
Phaeodactylum tricornutum	+	(in progress)	fcp	cat	ble	25
				luc	ble	26
				Aequorin		37
				EGFP, uidA	nat, sat1, nptII	29
				Glut1		54
			ca1			34
Thalassiosira pseudonana	-	+	-	-	-	18
Thalassiosira weissflogii	+	-	fcp	uidA	ble	26

Information for individual species is given only once in the list, even if promoters and genes were also used in subsequent publications. Abbreviations are: acc: acetyl CoA carboxylase; nptII: neomycin phosphotransferase; fcp: Fucoxanthin chlorophyll a/c-binding protein; frue: frustuline subunit e; Hup1 hexose transporter; ble: Zeocin/bleomycin binding protein; nr: nitrate reductase; cat: chloramphenicol actetyl transferase; luc: luciferine; EGFP: enhanced green fluorescent protein; uidA: b-Glucuronidase; nat: nourseothricin acetyltranferase; sat1: streptothricin acetyltranferase; Glut1: glucose transporter; ca1: carbonic anhydrase.

depending on whether the vector has integrated in a more or less active region of the chromosome. This effect allows the expression of problematic genes whose products might interfere with the cellular metabolism if highly expressed. In case of disruption of a functional endogenous gene there will always be a second allele present possibly complementing the knock-out. The design of the transformation vectors is very similar for *Phaeodactylum* and *Cylindrotheca* consisting of a bacterial standard vector for amplification in *E. coli* plus two eukaryotic promoters allowing expression of the selection marker as well as the gene of interest in the diatoms.[28,29] As in other algal systems, the highest expression rates are usually obtained when the promotor elements are derived from that organism in which they are supposed to be expressed. Heterologous promoters like the viral 35S-CaMV plant promoter[24] or the UHsp70 promoter from *Ustilago maydis*[25] were not functional in diatoms. So far mainly the strong light-induced promoters of the FCP proteins (Fucoxanthin chlorophyll a/c binding proteins)[25,26] and the constitutively expressed fruα3[28] protein have been utilized for driving the expression of selection markers as well as the respective genes of interest.[29] Recently Tanaka et al[34] described the successful utilization of the promoter of the carbonic anhydrase (ca1) of *Phaeodactylum*, while Poulsen and Kröger[27] introduced the first inducible promoter. They utilized the promoter of the nitrate reductase from *Cylindrotheca* which can be induced by the addition of nitrate to N-starved cells.[27] Interestingly the N-dependent regulation of nitrate reductase biosynthesis in *Cylindrotheca* is regulated on the translational rather than on the transcriptional level. Development of inducible promoters in diatoms is an essential step for future projects, because although it was shown that the light induced FCP promoter may result in the repression of green fluorescent protein (GFP) expression after 7 days in the dark and subsequent induction by 24 hours exposure of the cells to light,[35] light induction may not be generally applicable for studying diatom molecular biology as diatom growth is inhibited in the dark.

The selection of transgenic strains in most cases is based on the expression of the *sh ble* gene from the bacterium *Streptoallotheichus hindustanus*[36] which confers resistance to Zeocin. This antibiotic belongs to the bleomycin family. It enters the cells and causes cell death by intercalating into the DNA and cleaving it. Resistance is conferred by the *sh ble* protein binding to the antibiotic and preventing it from binding to DNA. Nourseothricine, strepthotricine and neomycin and its respective resistance genes have also been successfully utilized in diatoms.[29] Interestingly, *Phaeodactylum* has been demonstrated to withstand most of the other commonly used antibiotics and herbicides. Additionally the effectiveness of the various antibiotics may be significantly decreased in saline culture media,[25] therefore in some protocols *Phaeodactylum* strains were adapted to 50% or less seawater concentrations in order to reduce the amount of antibiotics needed for screening.

To analyse the expression of genes and to identify the subcellular location of proteins, several reporter genes like β-glucuronidase (GUS),[26,29] chloramphenicol acetyl transferase,[25] aequorin,[37] and a version of the GFP with an optimal codon usage for expression in human cells can be easily expressed in *Phaeodactylum* (EGFP).[38] The constructs are generally transferred as circular plasmids by particle bombardment using a biolistic device, resulting in comparably low efficiencies of 10^{-6} transformed cells/μg DNA. With both reporter gene and selection marker on the same plasmid, high rates of cotransformation of up to 60% can be achieved; however, when two plasmids are used expressing either GFP or CFP simultaneously similarly high rates of cells expressing both genes were obtained (Kilian and Kroth, unpublished), indicating that as soon as the foreign DNA gets into the cells and the cells are able to recover from the bombardment, there is a good chance that the DNA actually integrates into the nuclear genome.

In addition to expression of a foreign gene, another powerful benefit of molecular genetics is the possibility to knock-out or silence nuclear genes. To achieve this goal one approach can be the targeted disruption of a gene by insertion of a gene fragment or by replacement of the gene by a nonfunctional copy. Both approaches do not yet work in diatoms, probably because (i) vegetative diatom cells are diploid, thus if a gene should be knocked out there would still be another allele present, (ii) they require homologous recombination, a process by which the endogenous gene is replaced by a modified copy. Screening for such rare events requires higher transformation rates as are available now. Another approach to silence a gene is the expression of antisense RNA or RNAi resulting in a targeted destruction of the mRNA of the respective gene.[39] We recently found that antisense inhibition is feasible in *Phaeodactylum tricornutum*, but further work has to be done to understand the underlying mechanisms in diatoms (Materna and Kroth, unpublished results).

So far there is no publication available about plastid transformation systems for diatoms. This would allow the analysis of the genes of various proteins involved in photosynthesis. Similar to prokaryotic genomes homologous recombination has been demonstrated to occur in plastids of various organisms including green algae and land plants.[40,41] First experiments based on the transformation using resistance cassettes against antibiotics as well as on modification of plastid genes resulted in transformants that so far were only temporarily resistant (Materna and Kroth, unpublished results).

Biochemistry of Diatoms and Technological Applications

Diatoms have been useful in the past for various purposes like forensic applications, oil exploration and geological surveys (see ref. 42 for a detailed review). In addition diatoms served as biological monitors as there are a large number of ecologically sensitive species. On the other hand blooms of toxic and harmful diatoms may pose a growing problem for fisheries, aquaculture and public health,[43] thus understanding the biology of diatoms may prevent such blooms or at least may help to develop early alarm systems which may partially be based on genetic methods.[44] In addition to the potential utilization of diatom oils and frustules (see below), there are other diatom products that might benefit from genetic engineering. Diatomite is essentially used as abrasive material and for filtration.[10] Optimization of structure and porosity of the frustules might change its properties. Embedding frustules in metal-film membranes might result in gas selective membranes for medical applications.[45] As some diatoms are growing fast they could also be used as bioreactors for overexpression of proteins. Especially the plastid might be a useful target. It was shown that genetically modified tobacco chloroplasts may overexpress foreign proteins in such large amounts that the respective proteins represent up to 45 % of the total plastid protein.[46] A recent example for protein overexpression in algal plastids is the production and assembly of fully active antibodies.[47] Another example for protein expression in algal chloroplasts is the production of the foot-and-mouth-disease virus VP1 protein in *Chlamydomonas*,[47,48] demonstrating that algae may also be used as a mucosal vaccine source in future. These practical applications will be exposed in detail in Chapters 8 and 11.

Synthesis of Fatty Acids

A variety of algal cells contain oil droplets initially leading to the idea that they could be utilized as an alternative source of biofuels.[49] Dunahay et al[24,49] attempted to enhance lipid synthesis in the diatom *Cyclotella* by reintroducing multiple copies of the diatom's own acetyl CoA carboxylase gene. But, since photoautotrophic algae also consume large amounts of energy in the form of light in order to grow and produce oils (see below), such approaches did not result in commercially viable products. However, characterization of algal lipids resulted in the finding that various algae are able to produce long chain poly-unsaturated fatty acids (LCPUFAs). The most prominent ones for human health are docosahexaenoic acid (DHA) and eicosapentanoic acid (EPA).[50] DHA is a six-fold unsaturated, EPA a five-fold unsaturated fatty acid. One of the reasons why DHA plays an important role as a food additive is that humans contain about 20% of DHA in the grey matter of the brain and also in large amounts in the retina, but are not able to produce this fatty acid.[52] The common source for commercially produced DHA and EPA is fish oil or animal meat. Various algae have been found to produce LCPUFAs, including diatoms, chrysophytes, cryptophytes, and dinoflagellates.[50] The advantage of harvesting LCPUFAs from algae is mainly the higher purity of these oils because of lower amounts of contaminating oils. However, these parameters can greatly depend on species and culture conditions.[50] For instance, in the diatom ***Phaeodactylum tricornutum*** EPA may accumulate to up to 30% of the total amount of fatty acids. Recently genes for two desaturases have been identified in Phaeodactylum.[52] When expressed in yeast together with a heterologous elongase, detectable amounts of EPA accumulated within the yeast cells.

Unfortunately, most diatoms are obligatory photoautotrophs limiting the commercial impact because of the costs for illuminating the algal cultures. Even supplied with organic nutrients *Phaeodactylum* needs a minimum amount of light to sustain growth.[53] Therefore it was a great advancement when Zaslawskaia et al succeeded in the trophic conversion of the phototrophic *Phaeodactylum* by the introduction of a gene encoding the human glucose transporter into the diatom genome.[54] Expression and targeting of the protein into the plasma membrane of *Phaeodactylum* enabled the cells to grow heterotrophically in the dark in culture medium containing glucose. This clearly demonstrates that it might be possible to grow some diatoms heterotrophically as long as external carbohydrates can enter the cells. The possibility to turn obligate photoautotrophic microalgae genetically into heterotrophic cells in combination with the expression or inhibition of enzymes involved in the fatty acid metabolism might open up a wide range of commercial applications in the future.

Biomineralization

Different algae have acquired the ability to use inorganic ions to fabricate a cell wall that provides an effective mechanical protection.[55] The ability of diatoms to generate highly structured shells made of amorphous, hydrated SiO_2 is unmatched by modern technology. The two extracellular shells are joined by girdle bands, which are also made of silica. Silica is taken up into the cells by silicic acid transporters named SIT, which represent $NH_4^+/Si(OH)_4$ transporters that show no sequence homologies to other transporter proteins.[56] New frustules have to be synthesized before duplication of the cells. Therefore silica is condensated in special vesicles termed silica deposition vesicles (SDV) which are deposited near the plasma membrane of the dividing cells, followed by exocytosis of the whole cell wall.[57] For a long time it was completely unclear how the ornate structures are formed in such a highly ordered way. Pioneering work by Kröger et al discovered proteins termed silaffins and long chain polyamines to be involved in silica biogenesis.[58-60] These authors were also the first to report the incorporation of a foreign protein into the cell wall of *Cylindrotheca fusiformis* by genetic methods.[28] It was shown that both components (silaffins and polyamines) are able to accelerate silica formation from silicic acid in vitro and to influence the structure of the resulting silica.[61,62] Silaffins and other cell wall proteins like frustulins[63] and pleuralins[58] are synthesized as precursor forms which contain signal peptides for cotranslational transport into the endoplasmic reticulum. It is still unclear how the actual silica pattern is achieved, but several theoretical models have been proposed, including a model in which repeated phase separation events during wall biogenesis are assumed to produce self-similar silica patterns in smaller and smaller scales.[64] Biogenic silica can be highly interesting for future nanotechnology, as in contrast to industrial silica the structures are formed at moderate pH values, temperature and pressure and can be scaled up by simply enlarging the algal cultures.[65] Silica particles could either be used directly for incorporation into micro/nanodevices or they could be used as a scaffold to produce nanoparticles from other materials.[66] The fact that the structure of diatom frustules is species-specific clearly demonstrates that it is genetically controlled, thus it should be possible to influence or even control this process. However, the aim of producing "designed" silica particles in algae is still a far way off. We still do not know all the genes involved in silica biogenesis and, more strikingly, we do not know how the formation is actually controlled. Thus for future applications it will be inevitable to improve genetic methods to be able to identify these components and to characterize identified genes by modification, knock-out, and complementation in order to understand the process of biomineralization in diatoms.

Concluding Remarks

In contrast to other microalgae, diatoms so far have been used only for a few biotechnological applications. The main reason for the generally low competitiveness of algal products is the costs of photoautotrophic cultivation compared to other sources and the poor abilities to genetically modify them. Actually only ***Phaeodactylum*** is routinely transformed, which is not a typical diatom with respect to its poorly silicified cell walls. However, recent advances in algal genomics including the availability of 5′and 3′ noncoding sequences to utilize new promoters for transformation together with the increasing knowledge about diatom biology in general will surely result in various new options for the genetic modifications in the near future and increase the chances to develop new biotechnological applications and commercially viable products.

Acknowledgements

I am grateful for support by the University of Konstanz and for grants of the Deutsche Forschungsgemeinschaft (DFG) and the European community (MARGENES, contract QLRT-2001-01226) to PGK. Furthermore I would like to thank Prof. Dr. Kurt Mendgen and Rahul Bahulikar (University of Konstanz) for providing microscopical images of diatoms.

References

1. Lind JL, Heimann EA, van Vliet C et al. Substratum adhesion and gliding in a diatom are mediated by extracellular proteoglycans. Planta 1997; 203:213-221.
2. Lee RE. Phycology. Cambridge University Press, 1989.
3. Kooistra WHCF, De Stefano M, Mann DG et al. The phylogeny of the diatoms. Progr Mol Subcell Biol 2003; 33:59-97.
4. Norton TA, Melkonian M, Andersen RA. Algal biodiversity. Phycologia 1996; 35:308-326.
5. Round FE, Crawford RM, Mann DG. The diatoms: Biology and morphology of the genera. Cambridge University Press, 2005.
6. Chepurnov VA, Mann DG, Sabbe K et al. Experimental studies on sexual reproduction in diatoms. Int Rev Cytol 2004; 237:91-154.
7. Medlin LK, Kooistra WHCF, Gersonde R et al. Evolution of the diatoms (Bacillariophyta). II. Nuclear-Encoded small-subunit rRNA sequence comparisons confirm a paraphyletic origin for te centric diatoms. Mol Biol Evol 1996; 13:67-75.
8. Drum RW, Gordon R. Star Trek replicators and diatom nanotechnology. Trends Biotechnol 2003; 21:325-328.
9. Damste JS, Muyzer G, Abbas B et al. The rise of the rhizosolenid diatoms. Science 2004; 304:584-587.
10. Harwood DM. Diatomite. In: Stoermer EF, Smol JP, eds. The Diatoms: Applications for the Environmental and Earth Sciences. Cambridge University Press, 1999:436-443.
11. Falkowski PG, Barber RT, Smetacek V. Biogeochemical controls and feedbacks on ocean primary production. Science 1998; 281:200-205.
12. Bidle KD, Azam F. Accelerated dissolution of diatom silica by marine bacterial assemblages. Nature 1999; 397:508-512.
13. Falkowski PG, Katz ME, Knoll AH et al. The evolution of modern eukaryotic phytoplankton. Science 2004; 305:354-360.
14. Reinfelder JR, Milligan AJ, MorelFM. The role of the C4 pathway in carbon accumulation and fixation in a marine diatom. Plant Physiol 2004; 135:2106-2111.
15. Delwiche CF, Palmer JD. The origin of plastids and their spread via secondary symbiosis. Plant Syst Evol 1997; 11:53-86.
16. Cavalier-Smith T. Genomic reduction and evolution of novel genetic membranes and protein-targeting machinery in eukaryote-eukaryote chimaeras (meta-algae). Philos Trans R Soc Lond B Biol Sci 2003; 358:109-134.
17. Gibbs SP. The chloroplast endoplasmic reticulum: Structure, function, and evolutionary significance. Int Rev Cytol 1981; 72:49-99.
18. Armbrust EV, Berges JA, Bowler C et al. The genome of the diatom Thalassiosira pseudonana: Ecology, evolution, and metabolism. Science 2004; 306:79-86.
19. Edlund MB, Stoermer EF. Ecological, Evolutionary, and systematic significance of diatom histories. J Phycol 1997; 33:897-918.

20. Maheswari U, Montsant A, Goll J et al. The diatom EST database. Nucleic Acids Res 2005; 33(Database Issue):D344-D347.
21. Harris, EH. Chlamydomonas as model organism. Annu Rev Plant Physiol Plant Mol Biol 2001; 52:363-406.
22. Randolph-Anderson BL, Boynton JE, Gillham NW et al. Further characterization of the respiratory deficient dum-1 mutation of Chlamydomonas reinhardtii and its use as a recipient for mitochondrial transformation. Mol Gen Genet 1993; 236:235-244.
23. Boynton JE, Gillham NW. Genetics and transformation of mitochondria in the green alga Chlamydomonas. Methods Enzymol 1996; 264:279-96, (279-296).
24. Dunahay TG, Jarvis EE, Roessler PG. Genetic transformation of the diatoms Cyclotella cryptica and Navicula saprophila. J Phycol 1995; 31:1004-1012.
25. Apt KE, Kroth-Pancic PG, Grossman AR. Stable nuclear transformation of the diatom Phaeodactylum tricornutum. Mol Gen Genet 1996; 252:572-579.
26. Falciatore A, Casotti R, Leblanc C et al. Transformation of nonselectable reporter genes in marine diatoms. Marine Biotechnology 1999; 1:239-251.
27. Poulsen N, Kröger N. A new molecular tool for transgenic diatoms: Control of mRNA and protein biosynthesis by an inducible promoter-terminator cassette. FEBS J 2005; 272:3413-3423.
28. Fischer H, Robl I, Sumper M et al. Targeting and covalent modification of cell wall and membrane proteins heterologously expressed in the diatom Cylindrotheca fusformis. J Phycol 1999; 35:113-120.
29. Zaslavskaia LA, Lippmeier JC, Kroth PG et al. Transformation of the diatom Phaeodactylum tricornutum (Bacillariophyceae) with a variety of selectable marker and reporter genes. J Phycol 2000; 36:379-386.
30. Sanford JC, Smith FD, Russell JA. Optimizing the biolistic process for different biological applications. Methods Enzymol 1993; 217:483-509:483-509.
31. Kindle KL. High-frequency nuclear transformation of Chlamydomonas reinhardtii. Proc Natl Acad Sci USA 1990; 87:1228-1232.
32. Dunahay TG. Transformation of Chlamydomonas reinhardtii with silicon carbide whiskers. Biotech 1993; 15:452-460.
33. Shimogawara K, Fujiwara S, Grossman A et al. High-efficiency transformation of Chlamydomonas reinhardtii by electroporation. Genetics 1998; 148:1821-1828.
34. Tanaka Y, Nakatsuma D, Harada H et al. Localization of soluble β-carbonic anhydrase in the marine diatom phaeodactylum tricornutum. Sorting to the Chloroplast and Cluster Formation on the Girdle Lamellae. Plant Physiology 2005; 138:207-217.
35. Kilian O, Kroth PG. Identification and characterization of a new conserved motif within the presequence of proteins targeted into complex diatom plastids. Plant J 2005; 41:175-183.
36. Drocourt D, Calmels T, Reynes JP et al. Cassettes of the Streptoalloteichus hindustanus ble gene for transformation of lower and higher eukaryotes to phleomycin resistance. Nucleic Acids Res 1990; 18:4009.
37. Falciatore A, d'Alcala MR, Croot P et al. Perception of environmental signal by a marine diatom. Science 2000; 288:2363-2366.
38. Apt KE, Zaslavkaia L, Lippmeier JC et al. In vivo characterization of diatom multipartite plastid targeting signals. J Cell Sci 2002; 115:4061-4069.
39. van Dijk K, Marley KE, Jeong BR et al. Monomethyl histone H3 lysine 4 as an epigenetic mark for silenced euchromatin in Chlamydomonas. Plant Cell 2005; 17:2439-2453.
40. Bateman JM, Purton S. Tools for chloroplast transformation in Chlamydomonas: Expression vectors and a new dominant selectable marker. Mol Gen Genet 2000; 263:404-410.
41. Bogorad L. Engineering chloroplasts: An alternative site for foreign genes, proteins, reactions and products. Trends In Biotechnology 2000; 18:257-263.
42. Stoermer EF, Smol JP. Applications and uses of diatoms. In: Stoermer EF, Smol JP, eds. The diatoms: Applications for the environmental and earth Sciences. Cambridge University Press, 1999:3-8.
43. Fryxell GA, Villac MC. Toxic and harmful marine diatoms. In: Stoermer EF, Smol JP, eds. The Diatoms: Applications for the Environmental and Earth Sciences. Cambridge University Press, 1999:419-428.
44. Groben R, Medlin L. In situ hybridization of phytoplankton using fluorescently labeled rRNA probes. Methods Enzymol 2005; 395:299-310.
45. Wee KM, Rogers TN, Altan BS et al. Engineering and medical applications of diatoms. J Nanosci Nanotechnol 2005; 5:88-91.
46. De Cosa B, Moar W, Lee SB et al. Overexpression of the Bt cry2Aa2 operon in chloroplasts leads to formation of insecticidal crystals. Nat Biotechnol 2001; 19:71-74.

47. Mayfield SP, Franklin SE, Lerner RA. Expression and assembly of a fully active antibody in algae. Proc Natl Acad Sci USA 2003; 100:438-442.
48. Sun M, Qian K, Su N et al. Foot-and-mouth disease virus VP1 protein fused with cholera toxin B subunit expressed in Chlamydomonas reinhardtii chloroplast. Biotechnol Lett 2003; 25:1087-1092.
49. Dunahay TG, Jarvis EE, Dais SS et al. Manipulation of microalgal lipid production using genetic engineering. Appl Biochem Biotech 1996; 57/58:223-231.
50. Tonon T, Harvey D, Larson TR et al. A long chain polyunsaturated fatty acid production and partitioning to triacylglycerols in four microalgae. Phytochemistry 2002; 61:15-24.
51. Apt KE, Behrens PW. Commercial developments in microalgal biotechnology. J Phycol 1999; 35:215-226.
52. Domergue F, Spiekermann P, Lerchl J et al. New insight into phaeodactylum tricornutum fatty acid metabolism. Cloning and Functional Characterization of Plastidial and Microsomal Delta12-Fatty Acid Desaturases. Plant Physiol 2003; 131:1648-1660.
53. Fernandez Sevilla JM, Ceron Garcia MC, Sanchez Miron A et al. Pilot-plant-scale outdoor mixotrophic cultures of Phaeodactylum tricornutum using glycerol in vertical bubble column and airlift photobioreactors: Studies in fed-batch mode. Biotechnol Prog 2004; 20:728-736.
54. Zaslavskaia LA, Lippmeier JC, Shih C et al. Trophic obligate conversion of an photoautotrophic organism through metabolic engineering. Science 2001; 292:2073-2075.
55. Hamm CE, Merkel R, Springer O et al. Architecture and material properties of diatom shells provide effective mechanical protection. Nature 2003; 421:841-843.
56. Hildebrand M, Volcani BE, Gassmann W et al. A gene family of silicon transporters. Nature 1997; 385:688-689.
57. Pickett-Heaps J, Schmid AMM, Edgar LA. The cell biology of diatom valve formation. Progr Phycol Res 1990; 7:1-168.
58. Kröger N, Wetherbee R. Pleuralins are involved in Theca differentiation bin the diatom Cylindrotheca fusiformis. Protist 2000; 151:263-273.
59. Poulsen N, Sumper M, Kröger N. Biosilica formation in diatoms: Characterization of native silaffin-2 and its role in silica morphogenesis. Proc Nat Acad Sci 2003; 100:12075-12080.
60. Sumper M, Brunner E, Lehmann G. Biomineralization in diatoms: Characterization of novel polyamines associated with silica. FEBS Lett 2005; 579:3765-3769.
61. Kröger N, Deutzmann R, Bergsdorf C et al. Species-specific polyamines from diatoms control silica morphology. Proc Natl Acad Sci USA 2000; 97:14133-14138.
62. Kröger, Lorenz S, Brunner E et al. Self-assembly of highly phosphorylated silaffins and their function in biosilica morphogenesis. Science 2002; 298:584-586.
63. Kröger N, Bergsdorf C, Sumper M. Frustulins: Domain conservation in a protein family associated with diatom cell walls. Eur J Biochem 1996; 239:259-264.
64. Sumper M. A phase separation model for the nanopatterning of diatom biosilica. Science 2002; 295:2430-2433.
65. Hildebrand M. The prospects of manipulating diatom silica nanostructure. J Nanosci Nanotechnol 2005; 5:146-157.
66. Losic D, Mitchell JG, Voelcker NH. Complex gold nanostructures derived by templating from diatom frustules. Chem Commun (Camb) 2005; 4905-4907.

Chapter 4

Tools and Techniques for Chloroplast Transformation of *Chlamydomonas*

Saul Purton*

Abstract

The chloroplast organelle of plant and algal cells contains its own genetic system with a genome of a hundred or so genes. Stable transformation of the chloroplast was first achieved in 1988, using the newly developed biolistic method of DNA delivery to introduce cloned DNA into the genome of the green unicellular alga *Chlamydomonas reinhardtii*. Since that time there have been significant developments in chloroplast genetic engineering using this versatile organism, and it is probable that the next few years will see increasing interest in commercial applications whereby high-value therapeutic proteins and other recombinant products are synthesized in the *Chlamydomonas* chloroplast. In this chapter I review the basic methodology of chloroplast transformation, the current techniques and applications, and the future possibilities for using the *Chlamydomonas* chloroplast as a green organelle factory.

Introduction

The Chloroplast Genome

It is now generally accepted that the chloroplast organelle of plant and algal cells evolved from an oxygenic photosynthetic bacterium that established an endosymbiosis within a nonphotosynthetic eukaryotic host cell over one billion years ago. Since that time, the genome of this incarcerated bacterium has undergone a significant reduction in complexity, either through gene loss or gene transfer to the host nucleus, such that modern-day chloroplasts possess multiple copies of a small (120-200 kb) circular genome (or 'plastome') comprising some 100-250 genes. The majority of these genes encoded either components of the chloroplast's transcription-translation apparatus, or core components of the photosynthetic apparatus. Not only are these gene products homologous to those found in modern-day cyanobacteria, but the arrangement and expression of the genes also reflect the chloroplast's prokaryotic ancestry. Many genes are arranged in cotranscribed operons and the RNA polymerase and the 70S ribosome of the chloroplast are eubacterial rather than eukaryotic in nature.[1] The small size of the plastome, its relative simplicity and the prokaryotic nature of chloroplast gene expression make the chloroplast an attractive target for genetic engineering.

The Chloroplast as a Sub-Cellular Factory

The chloroplast compartment is the site of a number of important biosynthetic pathways and can also serve as a storage organelle. It is able to accumulate significant quantities of both

*Saul Purton—Algal Research Group, Department of Biology, University College London, Gower Street, London, WC1E 6BT, U.K. Email: s.purton@ucl.ac.uk

Transgenic Microalgae as Green Cell Factories, edited by Rosa León, Aurora Galván and Emilio Fernández.

soluble proteins and intrinsic membrane proteins (e.g., the rubisco enzyme and the core complexes of the photosynthetic apparatus, respectively), as well as large amounts of other macromolecules including chlorophylls, carotenoids, starch and lipids. The chloroplast is therefore an obvious cellular location for the synthesis and accumulation of valuable recombinant products.[2] Expressing the genes for these recombinant products in the chloroplast, rather than inserting the genes into the nucleus such that the protein is targeted into the chloroplast, is an attractive strategy for several reasons.[3] First, each chloroplast contains as many as a hundred copies of the plastome with genes encoding components of the photosynthetic apparatus expressed at a high level. Consequently, the copy number of a foreign gene (or 'transgene') inserted into the plastome is amplified significantly compared to the same gene inserted into the nuclear genome, and when fused to appropriate cis elements from these photosynthetic genes, the transgene has the potential for high-level expression. Second, chloroplast transformation studies in both algae and plants have shown that DNA integration occurs almost exclusively through homologous recombination. Not only does this allow precise and predictable manipulations of the plastome itself such that specific site-directed changes can be introduced into any chloroplast gene, but foreign DNA can be inserted at any chosen location within the plastome. As a consequence, the genetic engineering of the chloroplast avoids the problems of 'position effects' that are inherent with the random insertion of genes into the nuclear genome, and which affect significantly the level of transgene expression in the nucleus (see Chapter 1). Screening of large numbers of transformant lines is therefore not necessary when inserting transgenes into the chloroplast. Furthermore, integration via homologous recombination avoids the acquisition of undesirable *E. coli* vector sequences within the plastome. Third, studies to-date indicate that transgenes expressed in the chloroplast are not subject to the transcriptional and post-transcriptional gene silencing processes often observed with nuclear transgene expression. Fourth, multiple transgenes can be expressed as a single operon, allowing the coordinated synthesis of enzymes for a particular biosynthetic pathway or subunits of a protein complex. Fifth, the uniparental inheritance of chloroplast genes in many plant and algal species provides a mechanism for transgene containment.

Given these many advantages, it is perhaps not surprising that the manipulation of the plastome of plants and algae for commercial production of recombinant products—often referred to as "transplastomics"—is an increasingly active field.[3,4] A number of recent studies using tobacco chloroplasts have shown that remarkable yields of recombinant protein can be obtained; in some cases the protein accumulating to >25% of the total soluble protein (TSP) within the plant leaves.[3]

The Value of Chlamydomonas reinhardtii

The green unicellular alga *Chlamydomonas reinhardtii* occupies a special place in transgenic chloroplast research. The first demonstration of stable chloroplast transformation was achieved in 1988 using this alga,[5] and three years later the first recombinant protein produced in a chloroplast was by expression of a bacterial gene in the *Chlamydomonas* organelle.[6] Extensive reverse-genetic studies using *Chlamydomonas* over the past 18 years has provided significant insights into chloroplast gene function and expression, and a number of techniques first developed for plastome engineering in *Chlamydomonas* have been adapted for use in tobacco and other higher plants.[3] These include, the use of bacterial markers such as *aadA* and *aphA-6*, the technique of marker recycling and the use of plastome deletion mutants as recipient strains, as discussed below.

Whilst the levels of recombinant protein obtained in transplastomic *Chlamydomonas* have yet to match those obtained with tobacco (Table 1), this freshwater alga has several attributes that make it particularly attractive as a green cell factory. These are discussed in more detail by Fletcher et al in chapter 8, but to summarize: (1) *Chlamydomonas* chloroplast transformation is a simple and well-established technique that can generate transformant lines within 4-6 weeks. (2) The organism is a single cell with a single chloroplast, so any culture is a homogenous

Table 1. Foreign genes expressed in the* C. reinhardtii *chloroplast

Gene	Source Organism	Gene Product	Yield of Recombinant Protein (as % TSP)	Ref.
aadA	eubacteria (R plasmid)	Aminoglycoside adenyl transferase	nd	6
uidA	*Escherichia coli*	β-glucoronidase	nd	27
recA	*Escherichia coli*	RecA protein	nd	18
uidA	*Escherichia coli*	β-glucoronidase	0.08%	43
rluc	*Renilla reniformis*	Luciferase	nd	44
aphA6	*Acinetobacter baumannii*	Aminoglycoside phosphotransferase	nd	23
gfp cDNA	*Aequora victoria*	Green fluorescent protein	~0.05%	28
GFPct	Synthetic gene	Green fluorescent protein	0.5%	29
gds	*Sulfolobus acidocaldarius*	Geranylgeranyl pyrophosphate synthase	0.1%	34
HSV8-lsc	Synthetic gene	Large single-chain antibody	0.5%	8
FMDV.VP1	foot-and-mouth disease virus	Viral protein 1	3%	33
luxCt	Synthetic gene	Bacterial luciferase	nd	30
apcA and *apcB*	*Spirulina maxima*	α and β subunits of allophycocyanin	2-3%	36
nifH	*Klebsiella pneumoniae*	Nitrogenase subunit	nd	35

nd: not determined

collection of cells with all cells actively expressing the transgene. (3) *Chlamydomonas* has a short generation time of less than eight hours under optimum conditions, allowing the rapid production of biomass. (4) Importantly, *Chlamydomonas* can dispense completely with photosynthetic function and can be grown in darkness when supplied with acetate as an exogenous fixed carbon source. This allows the mass cultivation of transgenic lines in traditional fermentors under controlled, sterile and axenic culture conditions. Not only does this provide containment of the genetically modified organism, but also ensures that the production of pharmaceutical products can meet the strict regulations laid down by regulatory bodies such as the U.S. Food and Drug Administration. (5) *Chlamydomonas* is classified as a *GRAS* ('generally regarded as safe') organism with no known viral or bacterial pathogens, and appears to lack the endotoxins that complicate recombinant protein production using *Escherichia coli* and other bacterial platforms. (6) Our own studies have indicated that the 'escape' of chloroplast DNA sequences to the nuclear genome of *Chlamydomonas* is several orders of magnitude less frequent than that observed in tobacco, thereby increasing the containment level of transgenes in the chloroplast.[7]

The potential of the *Chlamydomonas* chloroplast as an expression platform has been elegantly illustrated by the recent work of Mayfield and colleagues who were able to express a fully active human monoclonal antibody in the alga.[8] In the following sections I review the various tools and techniques for chloroplast transformation of *Chlamydomonas* and illustrate its application in both basic research and more recent forays into applied areas of research. For more detailed discussion on the early development of chloroplast transformation in this alga, the reader is directed to three excellent reviews.[9-11]

Delivery of DNA into the Chloroplast Compartment

The most efficient and reliable method for introducing DNA into the chloroplast is microparticle bombardment. This biological-ballistic, or 'biolistic' process involves placing a nutrient agar plate carrying sections of plant tissue or a lawn of algal cells into a vacuum chamber, and bombarding the plate with DNA-coated gold or tungsten microparticles.[12] The kinetic energy of these particles is sufficient to penetrate the cell wall, the plasma membrane and the two membranes surrounding the chloroplast, and deliver multiple copies of the transforming DNA into the organelle. Early biolistic devices used for chloroplast transformation were home-built devices that used gunpowder charges to accelerate the microparticles.[13,14] Nowadays, most *Chlamydomonas* laboratories use the commercial helium-powered device marketed by Bio-Rad. Helium has the advantage of being a cleaner and more controllable propellant, and this together with several other design improvements has resulted in significantly increased transformation rates. The microparticles used in the bombardment process are made of either tungsten or gold, with gold being the more expensive but preferred material for most workers. This preference is because of: (i) the gold particles more uniform size; (ii) the biologically inert nature of gold as compared to tungsten, which can be toxic or mildly mutagenic to some cell types, and (iii) the tendency of tungsten to catalyze the slow degradation of DNA bound to it.

Several workers have investigated alternatives to biolistics as methods for delivering DNA into the *Chlamydomonas* chloroplast. Kindle and colleagues[15] showed that chloroplast transformation can be achieved simply by agitating a suspension of cells, DNA and glass beads using a laboratory vortex. Alternatively, electroporation can be used to generate chloroplast transformants (B. Sears, personal communication). Although the transformation rates using these methods are lower than those obtained with biolistics, and both methods require that a cell-wall-deficient strain is used as the recipient, the much lower costs of the equipment needed make them attractive alternatives to biolistics.

Integration of Transforming DNA

An important finding that came from the first report on chloroplast transformation was that transforming DNA carrying cloned chloroplast sequences integrated into the plastome via homologous recombination between these sequences and the corresponding endogenous sequences.[5] As illustrated in Figure 1, this allows one to carry out precise and premeditated manipulations of the plastome – a process that Rochaix termed 'chloroplast DNA surgery'.[16] In theory, any nonessential endogenous gene can be deleted or modified, with modifications ranging from one or more site-directed changes to the gene sequence, to replacement of promoter or untranslated regions (UTRs), or insertion of novel DNA elements that may influence the gene's expression. Furthermore, foreign genes can be inserted at any chosen position, and in either orientation, within the plastome (Fig. 2). This is in marked contrast to nuclear transformation in *Chlamydomonas*, where transforming DNA integrates at apparently random genomic loci via nonhomologous recombination processes.[17] Consequently, precise engineering of nuclear genes of *Chlamydomonas* is not feasible, and successful expression of foreign genes introduced into the nuclear genome unpredictable (see also Chapters 1 and 7).

Stable integration of foreign DNA into the plastome requires sufficient homologous sequence flanking the DNA to allow two recombination events (the actual mechanism of DNA integration is not known, but is best envisaged as a simple double crossover event as shown in Figs. 1 and 2). What constituent 'sufficient homologous sequence' has not been rigorously examined, although a good rule of thumb is ~1 kb of homology on either side of the DNA to be introduced. This is clearly sufficient since several studies have shown that inter-molecular recombination between incoming DNA and the plastome, or intra-molecular recombination between direct repeats flanking DNA inserted into the plastome can occur with as little as 120

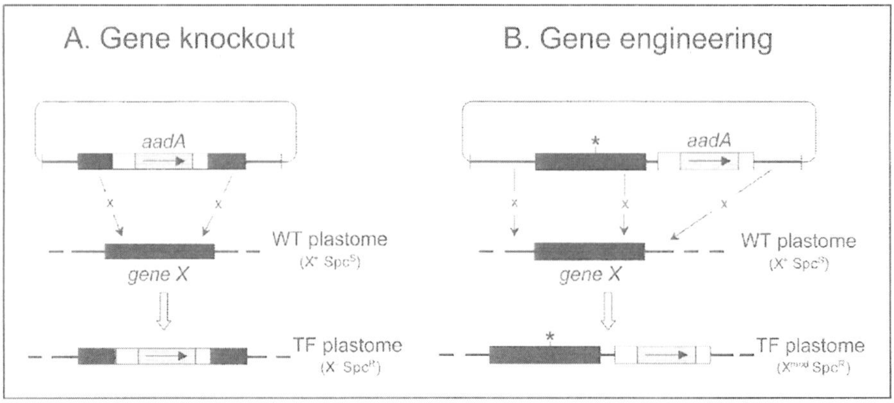

Figure 1. Reverse-genetic analysis of chloroplast gene X. Homologous recombination between chloroplast sequences (thick black lines) on the plasmid and the wildtype (WT) plastome allows disruption (A) or site-directed modifications (B) of the target gene using, for example, the *aadA* cassette[6] to select for spectinomycin (Spc) resistant transformants (TF). In the case of [B], the site-directed change may not be introduced if recombination occurs between the engineered change and *aadA* (broken arrow).

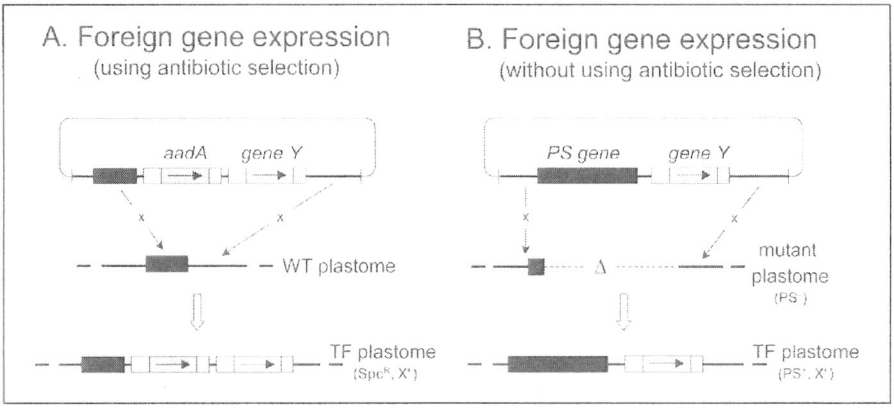

Figure 2. Two different strategies for introducing your foreign gene Y into the plastome. Gene Y fused to endogenous 5' and 3' sequences (open boxes) can be linked either to the *aadA* marker as shown in (A), allowing selection for spectinomycin-resistant transformants (TF), or to an endogenous gene required for photosynthesis (PS) as shown in (B), allowing selection for restored photosynthetic function.

bp and 216 bp, respectively of homologous sequence.[18,19] However, successful recombination using such short elements may be dependent on the sequences chosen.[11]

There are conflicting claims as to whether linear or circular DNA yields better transformation rates (see ref. 10), although any difference is marginal and since the researcher usually analyses only a few transformant colonies from the hundreds obtained in a typical transformation experiment, then most researchers choose the convenience of uncut plasmid DNA, as depicted in Figures 1 and 2.

If the transforming DNA is flanked on one side only with a homologous element then a single crossover takes place and results in the integration of the whole plasmid into the plastome with direct repeats of the element flanking the plasmid sequence. This integration is genetically

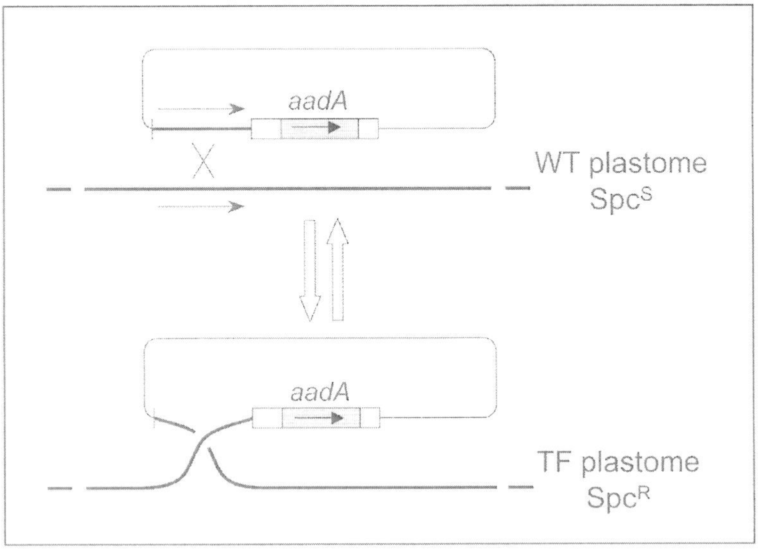

Figure 3. Transformation with a plasmid construct in which foreign DNA such as *aadA* is flanked on one side only with plastome sequence. This leads to a reversible transplastomic state in which the whole plasmid integrates into the plastome but is readily excised through recombination between the resulting direct repeats.

unstable since recombination between the repeats can result in the subsequent excision of the plasmid (Fig. 3). Consequently, the removal of the selective pressure for maintenance of the transforming DNA eventually results in the loss of the DNA since it is unlikely that the free plasmid will be replicated in the chloroplast. In theory, such an approach could be used as an alternative strategy to those developed by Fischer et al[20] (see below) for elimination of a selectable marker from the plastome.

Occasionally, unexpected recombination events can arise when the transforming plasmid contains homologous sequences in addition to the two flanking sequences—for example promoter elements or UTRs from endogenous genes that are used to drive expression of the selectable marker or other foreign genes. Intra- or intermolecular recombination between these sequences and the endogenous gene sequence can result in rearrangements or deletions.[11] Depending on the relative position and orientation of these two copies of the sequence, the result may be one of two states as explained in the next section: a heteroplasmic state in which there is a minor population of fragmented plastomes continuously being generated within the chloroplast, or a homoplasmic state where all plastome copies carry a deletion extending from the insertion site to the endogenous gene.[21]

Polyploidy and the Problems of Heteroplasmy

The plastome of *Chlamydomonas* is highly polyploid with approximately 50-80 copies present within the chloroplast. Transformation therefore invariably results in an initial heteroplasmic state in which only some copies of the plastome within an individual cell have been modified. Growth of transformant lines under selective conditions should eventually result in homoplasmic cells in which all copies of the plastome carry the selected change (Fig. 4). This homoplasmic state is attained typically by taking the lines through several rounds of single colony isolation on selective medium, and once homoplasmy is achieved the transgenic plastome is stable even in the absence of selection. In contrast, the initial heteroplasmic state will persist despite prolonged selective pressure if the engineered change disrupts an essential chloroplast gene. In this

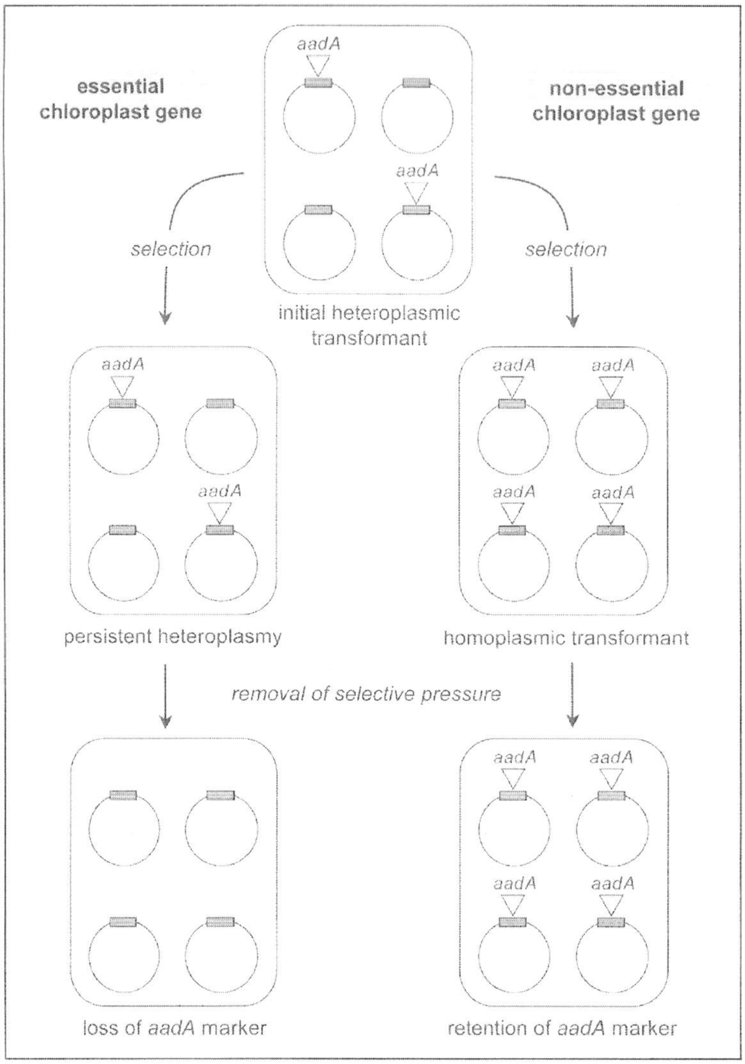

Figure 4. Two possible plastome outcomes to a chloroplast gene knockout experiment using *aadA* depending on whether or not the target gene (shown as a shaded box on the circular plastomes within a single chloroplast) is essential for cell survival.

case, the requirement for retention of the essential gene counters that for the selectable marker and both types of plastome are maintained. Subsequent removal of the selective pressure for the transgenic copy rapidly results in the loss of these plastomes from the chloroplast (Fig. 4).

Heteroplasmy can also prove problematic when recessive mutations are introduced into chloroplast genes, for example site-directed changes that affect the functioning of a photosynthetic gene. Since there is a natural selection against such mutations (unless transformant colonies are grown in the dark or in the presence of inhibitors of photosynthesis), then introducing these changes can prove challenging. Firstly, there is a tendency to recover transformants arising from recombination events that have incorporated the selectable marker into plastome copies, but not the linked site-directed change (see Fig. 1B). Secondly, the change may be successfully introduced into some copies of the plastome but these are subsequently lost during the rounds of selection since selection favors those plastome copies carrying only the selectable marker. Finally, 'copy correction' mechanisms (either intermolecular recombination or gene conversion) may act during the heteroplasmic state, repairing the mutated gene copies using wild-type copies of the gene. An understanding of these issues is important not just for reverse-genetic studies of chloroplast gene function, but also for the introduction of foreign genes into the plastome. It is likely that many foreign genes will have some level of detrimental effect on the biology of the chloroplast, and can therefore be considered as recessive mutations. Consequently, the same selective pressures will affect the recovery of homoplasmic lines carrying the foreign gene, although in this case copy-correction would involve repair of the locus containing the inserted DNA using copies of the 'empty' wild-type locus.

Two strategies have been developed to overcome these problems. The first involves the treatment of *Chlamydomonas* with 0.5 mM 5-fluordeoxyuridine (FdUrd) prior to transformation.[9] This nucleoside analog appears to be a specific inhibitor of chloroplast DNA replication, reducing the plastome copy number by as much as 10-fold. Since a microparticle entering the chloroplast following bombardment will carry multiple copies of the transforming DNA, then the use of FdUrd significantly increases the chances of the recessive mutation being introduced into all or most plastome copies, and homoplasmy being attained without loss or repair of the mutation. However, FdUrd is also a chemical mutagen, and can lead to unwanted mutations in the plastome. A second strategy, pioneered by Redding and colleagues[22] uses a chloroplast deletion mutant as the recipient strain, and neatly addresses all of the problems discussed above. By ensuring that the homologous DNA carried on the transforming plasmid spans the deletion within the mutant plastome, and that both recessive mutation and the selectable marker are located within the deleted region, then modified plastome copies in the initial heteroplasmic transformants will always have both the selectable marker and the recessive mutation (one such scenario is illustrated in Fig. 2B). Furthermore, no wild-type copies of the mutated locus are present in the chloroplast so elimination via copy correction is not possible.

Selection Strategies

Several different strategies have been developed for the selection of rare transformants amongst the large population of untransformed cells following bombardment of the algal lawn. In the early transformation experiments, selection was based on the rescue of nonphotosynthetic mutants carrying lesions in chloroplast genes encoding key photosynthetic components.[10] For example, the original experiment of Boynton et al involved the bombardment of a mutant line carrying a plastome deletion affecting the *atpB* gene with a plasmid carrying the wild-type gene.[5] Introduction of *atpB* into the plastome of the mutant restored photosynthetic function, and hence the ability to grow on a medium lacking a reduced carbon source. Using this approach, foreign genes can be targeted to a neutral, intergenic region downstream of *atpB* (Fig. 2B), or can be targeted to other loci by cotransformation using the *atpB* plasmid and a second plasmid in which the foreign gene is flanked with the appropriate homologous sequence. The main drawback of this selection strategy is that a specific deletion mutant is required as the recipient rather than any strain carrying a wild-type plastome. There are, however, two specific advantages: firstly, there is strong selection for transformant clones since a wild-type (i.e., photosynthetically competent and light tolerant) phenotype is being restored, and therefore transformants are easily generated and rapidly attain a homoplasmic state. Secondly, foreign DNA can be introduced into the plastome without the use of any bacterial antibiotic-resistance marker. This circumvents one of the major environmental concerns with transplastomic plants and algae - the presence within each cell of multiple copies of bacterial markers and the possibility of their horizontal spread to other organisms in the environment.

An alternative method that allows the use of wild-type recipient strains involves selection for antibiotic resistance using cloned *Chlamydomonas* plastome genes carrying dominant point mutations. For example, mutations that confer resistance to different antibiotics have been identified in genes encoding the 16S and 23S ribosomal RNAs, the ribosomal protein S12 and elongation factor Tu.[9] As before, the advantage of this strategy is that no bacterial marker is used. However, the disadvantages are that selection on these antibiotics inevitably yields a background of pseudo-transformants (i.e., spontaneous resistance mutants) amongst the transformant population, and the dominant mutations introduced into the chloroplast's translational apparatus may affect the efficiency of protein synthesis - especially when trying to over-express foreign genes.

The most widely used dominant marker is the '*aadA* cassette' developed by Goldschmidt-Clermont.[6] This marker comprises the coding sequence of the bacterial gene *aadA* fused to upstream and downstream elements from *Chlamydomonas* chloroplast genes, and confers resistance to the antibiotics, spectinomycin and streptomycin. More recently, my group has developed a similar marker based on the bacterial *aphA-6* gene that confers resistance to kanamycin and amikacin.[22] As illustrated in Figure 1, these portable cassettes can be targeted to any site on the plastome and used to disrupt or modify endogenous genes, or to introduce foreign genes. However, the limited availability of different dominant markers led Fischer et al to explore different ways of 'recycling' the *aadA* cassette so that serial manipulations of the plastome would be possible using the same marker.[20] By flanking the cassette with direct repeat elements or placing it within an essential gene, Fischer et al showed that the cassette was rapidly lost from the transformed plastome once spectinomycin selection was removed. As discussed above, these "use it, then lose it" strategies also overcome the issue of unwanted bacterial antibiotic markers in the resulting transplastomic lines.

Reverse-Genetic Studies of the *Chlamydomonas* Plastome

The development of a routine and facile method for manipulating the *Chlamydomonas* plastome has allowed a wide range of reverse-genetic studies that have provided fundamental insights into the expression of chloroplast genes and the functioning of their products.[24,25] Some selected examples are as follows: (1) gene disruption studies have identified genes encoding novel components of the photosynthetic complexes, such as the PetL subunit of the cytochrome b_6f complex, and the PsbT and PsbZ subunits of photosystem II.[24] (2) Site-directed changes and nucleotide insertions or deletions around the start codon of chloroplast genes have allowed a dissection of cis-acting sequences required for translation initiation.[26] (3) Site-directed changes to conserved residues of the core subunits of photosystem I (PSI) have identified those residues that modulate the properties of the redox cofactors involved in electron transfer and led to the finding that electron transfer is bi-directional in PSI.[22,25] (4) Reporter genes have been used to investigate the role of the 5' and 3' UTRs of chloroplast genes in mediating RNA stability and translation.[27,28] Furthermore, the recent development of new codon-optimized reporter genes encoding green fluorescent protein[29] or luciferase[30] now opens the door to even more sensitive and sophisticated gene expression studies. (5) Mutational studies of the cytochrome *f* subunit of the cytochrome b_6f complex has identified residues that constitute a translation repression motif involved in the assembly-mediated regulation of cytochrome *f* synthesis.[31] (6) The topology of proteins embedded in the thylakoid membrane has been investigated using *aadA* translational fusions, whereby fusions that place the AAD protein on the stromal side of the membrane yield high levels of spectinomycin resistance whereas those that place it on the luminal side give low resistance levels.[32]

Expression of Foreign Genes in the *Chlamydomonas* Chloroplast

The expression of foreign genes in the *Chlamydomonas* chloroplast is still in its infancy with only a dozen or so reports in the literature (Table 1). Most of these reports represent basic research aimed at developing molecular tools, and include the selectable markers and reporter genes discussed in the previous sections. In each case, the gene construct consists of a promoter, together with 5' and 3' UTRs from endogenous chloroplast genes, fused to coding sequence for the foreign gene. The coding sequence is either obtained directly from bacterial genes (in the case of *aadA*, *aphA-6* and *uidA*) or is synthesized de novo using the preferred codon usage observed for *Chlamydomonas* chloroplast genes. To-date there has been no reports of successful expression of foreign genes using synthetic promoters or those from other organisms (e.g., from bacterial genes or organellar genes of other eukaryotic species). There have been only a handful of papers describing applied research aimed at producing novel products or introducing novel biosynthetic pathways. As discussed in Chapter 8, Mayfield and colleagues have demonstrated the feasibility of producing soluble, active monoclonal antibodies in *Chlamydomonas*, and their research has illustrated the importance of codon optimization in maximizing transgene expression.[8] Importantly, their studies showed that the single-chain antibody folded correctly and dimerized via a disulfide bridge, but was not subject to unwanted post-translational modifications such as glycosylation within the chloroplast compartment. In a separate study, Sun et al have explored the possibility of vaccine production by expressing a fusion construct comprising the VP1 gene of foot-and-mouth virus fused to the gene for Cholera toxin B.[33] Again, initial studies suggest proper folding of the recombinant protein in the chloroplast.

Promising early work on the manipulation of chloroplast metabolic pathways in *Chlamydomonas* has been reported by several groups. Fukusaki et al successfully expressed an archeal gene encoding a thermostable version of the GGPP synthase enzyme involved in terpenoid biosynthesis,[34] and Cheng et al were able to functionally replace a chloroplast gene involved in chlorophyll biosynthesis with a related gene from *Klebsiella pneumoniae* encoding a component of the nitrogen-fixing enzyme, nitrogenase.[35] Finally, the possibility of expressing multiple foreign genes in *Chlamydomonas*, and thereby constructing novel biosynthetic pathways or multisubunit protein complexes, has been investigated by Su et al who were able to express a two gene operon encoding the alpha and beta subunits of the cyanobacterial light-harvesting protein, allophycocyanin.[36]

Future Prospects

There is currently much interest in transplastomics as a strategy for the commercial production of recombinant macromolecules.[4] Most of this interest has focused on cultivatable plants, not least because of the potential for large-scale production using established agricultural practices. Whilst fermentor-based culturing of green algae such as *Chlamydomonas* can never match field-based cultivation of plants such as tobacco in terms of chloroplast biomass, there is a possible niche market for *Chlamydomonas* in this emerging biotech industry. This niche encompasses high-value recombinant products intended for human use and includes pharmaceutical proteins (e.g., hormones, vaccines and antibodies), nutriceuticals (e.g., speciality carotenoids and long-chain polyunsaturated fatty acids) and other bioactive secondary metabolites.

However, there are several further developments that are required before *Chlamydomonas* can realize its potential. Key amongst these is a significant improvement in recombinant protein yield. Current yields are, at best, one or two percent of TSP (Table 1)—significantly below that of endogenous chloroplast proteins such as the large subunit of the rubisco enzyme or the D1 and D2 proteins of photosystem II. The accumulating mass of published data on chloroplast gene expression in *Chlamydomonas* is highlighting those factors that are important considerations for improving protein levels. Post-transcriptional steps seem to be more critical than gene copy number, transcription rates or transcript levels,[37] and the focus now is on translation initiation, polypeptide synthesis and protein stability. For the first of these, the focus is the cis elements within the 5' UTRs or coding regions of chloroplast genes and how these elements interact with nuclear-encoded trans-acting factors to mediate efficient translation initiation.[38] In terms of efficient translation, the work of Mayfield et al has shown that codon optimization can significantly improve the synthesis of recombinant proteins.[8,29,30] Finally, a clearer understanding of the targets and mechanisms of the different chloroplast proteases may provide opportunities for increased protein accumulation using protease-deficient strains.

In addition to maximizing recombinant protein levels, the commercial production of recombinant proteins in the *Chlamydomonas* chloroplast also requires the development of an inducible system for transgene expression. Currently, all foreign genes expressed in this alga's organelle are driven from endogenous promoters (e.g., those from *rrn16*, *atpA* or *rbcL*) and therefore transgene expression is "on" throughout all the growth phases. What is much more preferable is a system whereby expression is suppressed until late log phase allowing the over-accumulation of the recombinant protein in the cells only once sufficient biomass has been generated. Such an inducible system also allows the production of proteins that have toxic or detrimental effects on the growing cells. Several systems have been developed recently that allow the inducible expression of chloroplast transgenes in tobacco[39,40] and it should be feasible to develop similar strategies for *Chlamydomonas*.

Given such improvements, together with the relative ease with which unicellular algae such as *Chlamydomonas* can be grown heterotrophically on an industrial scale using existing fermentor-based technology,[41] and at minimal cost,[42] it is possible to imagine a future where this alga's chloroplast is a viable and commercially attractive platform for recombinant production, and is able to compete with existing microbial platforms based on bacteria or yeast.

Acknowledgements

Chlamydomonas research in my laboratory is supported by the U.K. Biotechnology and Biological Sciences Research Council and the Leverhulme Trust.

References

1. Sugiura M. The chloroplast genome. Essays Biochem 1995; 30:49-57.
2. Bogorad L. Engineering chloroplasts: An alternative site for foreign genes, reactions and products. Trends Biotechnol 2000; 18:257-263.
3. Maliga P. Plastid transformation in higher plants. Annu Rev Plant Biol 2004; 55:289-313.
4. Franklin SE, Mayfield SP. Prospects for molecular farming in the green alga Chlamydomonas. Curr Opin Plant Biol 2004; 7:159-165.
5. Boynton JE, Gillham NW, Harris EH et al. Chloroplast transformation in Chlamydomonas with high velocity microprojectiles. Science 1988; 240:1534-1538.
6. Goldschmidt-Clermont M. Transgenic expression of aminoglycoside adenine transferase in the chloroplast: A selectable marker for site-directed transformation of Chlamydomonas. Nucleic Acids Res 1991; 19:4083-4089.
7. Lister DL, Bateman JM, Purton S et al. DNA transfer from chloroplast to nucleus is much rarer in Chlamydomonas than in tobacco. Gene 2003; 316:33-38.
8. Mayfield SP, Franklin SE, Lerner RA. Expression and assembly of a fully active antibody in algae. Proc Natl Acad Sci USA 2003; 100:438-442.
9. Boynton JE, Gillham NW. Chloroplast transformation in Chlamydomonas. Methods Enzymol 1993; 217:510-536.
10. Erickson JM. Chloroplast transformation: Current results and future prospects. In: Ort DR, Yocum CF, eds. Oxygenic Photosynthesis: The Light Reactions. Dordrecht: Kluwer Academic Publishers, 1996:589-619.
11. Goldschmidt-Clermont M. Chloroplast transformation and reverse genetics. In: Rochaix JD, Goldschmidt-Clermont M, Merchant S, eds. The Molecular Biology of Chloroplasts and Mitochondria in Chlamydomonas. Dordrecht: Kluwer Academic Publishers, 1998:139-149.
12. Taylor NJ, Fauquet CM. Microparticle bombardment as a tool in plant science and agricultural biotechnology. DNA Cell Biol 2002; 21:963-977.
13. Sanford JC. The biolistic process. Trends Biotechnol 1998; 6:299-302.
14. Zumbrunn G, Schneider M, Rochaix JR. A simple particle gun for DNA-mediated cell transformation. Technique 1989; 1:204-216.
15. Kindle KL, Richards KL, Stern DB. Engineering the chloroplast genome: Techniques and capabilities for chloroplast transformation in Chlamydomonas reinhardtii. Proc Natl Acad Sci USA 1991; 88:1721-1725.
16. Rochaix JD. Chlamydomonas reinhardtii as the photosynthetic yeast. Annu Rev Genet 1995; 29:209-230.
17. Walker TL, Collet C, Purton S. Algal transgenics in the genomic era. J Phycol 2005; 41:1077-1093.
18. Cerutti H, Johnson AM, Boynton JE et al. Inhibition of chloroplast DNA recombination and repair by dominant negative mutants of Escherichia coli RecA. Mol Cell Biol 1995; 16:3003-3011.

19. Dauvillee D, Hilbig L, Preiss S et al. Minimal extent of sequence homology required for homologous recombination at the psbA locus in Chlamydomonas reinhardtii chloroplasts using PCR-generated DNA fragments. Photosynth Res 2004; 79:219-224.
20. Fischer N, Stampacchia O, Redding K et al. Selectable marker recycling in the chloroplast. Mol Gen Genet 1996; 251:373-380.
21. Künstner P, Guardiola A, Takahashi Y et al. A mutant strain of Chlamydomonas reinhardtii lacking the chloroplast photosystem II psbI gene grows photoautotrophically. J Biol Chem 1995; 270:9651-9654.
22. Guergova-Kuras M, Boudreaux B, Joliot A et al. Evidence for two active branches for electron transfer in photosystem I. Proc Natl Acad Sci USA 2001; 98:4437-4442.
23. Bateman JM, Purton S. Tools for chloroplast transformation in Chlamydomonas: Expression vectors and a new dominant selectable marker. Mol Gen Genet 2000; 263:404-410.
24. Rochaix JD. Functional analysis of plastid genes through chloroplast reverse genetics in Chlamydomonas. In: Larkum AWD, Douglas SE, Raven JA, eds. Photosynthesis in Algae. Dordrecht: Kluwer Academic Publishers, 2003:83-94.
25. Xiong L, Sayre RT. Engineering the chloroplast encoded proteins of Chlamydomonas. Photosynth Res 2004; 80:411-419.
26. Hauser CR, Gillham NW, Boynton JE. Regulation of chloroplast translation. In: Rochaix JD, Goldschmidt-Clermont M, Merchant S, eds. The Molecular Biology of Chloroplasts and Mitochondria in Chlamydomonas. Dordrecht: Kluwer Academic Publishers, 1998:197-217.
27. Sakamoto W, Kindle KL, Stern DB. In vivo analysis of Chlamydomonas chloroplast petD gene expression using stable transformation of beta-glucuronidase translational fusions. Proc Natl Acad Sci USA 1993; 90:497-501.
28. Komine Y, Kikis E, Schuster G et al. Evidence for in vivo modulation of chloroplast RNA stability by 3'-UTR homopolymeric tails in Chlamydomonas reinhardtii. Proc Natl Acad Sci USA 2002; 99:4085-4090.
29. Franklin S, Ngo B, Efuet E et al. Development of a GFP reporter gene for Chlamydomonas reinhardtii chloroplast. Plant J 2002; 30:733-744.
30. Mayfield SP, Schultz J. Development of a luciferase reporter gene, luxCt, for Chlamydomonas reinhardtii chloroplast. Plant J 2004; 37:449-458.
31. Choquet Y, Zito F, Wostrikoff K et al. Cytochrome f translation in Chlamydomonas chloroplast is autoregulated by its carboxyl-terminal domain. Plant Cell 2003; 15:1443-1454.
32. Franklin JL, Zhang J, Redding K. Use of aminoglycoside adenyltransferase translational fusions to determine topology of thylakoid membrane proteins. FEBS Lett 2003; 536:97-100.
33. Sun M, Qian K, Su N et al. Foot-and-mouth disease virus VP1 protein fused with cholera toxin B subunit expressed in Chlamydomonas reinhardtii chloroplast. Biotechnol Lett 2003; 25:1087-1092.
34. Fukusaki EI, Nishikawa T, Kato K et al. Introduction of the archaebacterial geranylgeranyl pyrophosphate synthase gene into Chlamydomonas reinhardtii chloroplast. J Biosci Bioeng 2003; 95:283-287.
35. Cheng Q, Day A, Dowson-Day M et al. The Klebsiella pneumoniae nitrogenase Fe protein gene (nifH) functionally substitutes for the chlL gene in Chlamydomonas reinhardtii. Biochem Biophys Res Comm 2005; 329:966-975.
36. Su ZL, Qian KX, Tan CP et al. Recombination and heterologous expression of allophycocyanin in the chloroplast of Chlamydomonas reinhardtii. Acta Biochimica et Biophysica Sinica 2005; 37:709-712.
37. Eberhard S, Drapier D, Wollman FA. Searching limiting steps in the expression of chloroplast-encoded proteins: Relations between gene copy number, transcription, transcript abundance and translation rate in the chloroplast of Chlamydomonas reinhardtii. Plant J 2002; 31:149-160.
38. Barnes D, Franklin S, Schultz et al. Contribution of 5'- and 3' untranslated regions of plastid mRNAs to the expression of Chlamydomonas reinhardtii chloroplast genes. Mol Gen Genomics 2005; 274:625-636.
39. Lössl A, Bohmert K, Harloff H et al. Inducible trans-activation of plastid transgenes: Expression of the R. eutropha phb operon in transplastomic tobacco. Plant Cell Physiol 2005; 46:1462-1471.
40. Mühlbauer SK, Koop HU. External control of transgene expression in tobacco plastids using the bacterial lac repressor. Plant J 2005; 43:941-946.
41. Apt KE, Behren PW. Commercial developments in microalgal biotechnology. J Phycol 1999; 35:215-226.
42. Franklin SE, Mayfield SP. Recent developments in the production of human therapeutic proteins in eukaryotic algae. Expert Opin Biol Ther 2005; 5:225-235.
43. Ishukura K, Takaoka Y, Kato K et al. Expression of a foreign gene in Chlamydomonas reinhardtii chloroplast. J Biosci Bioeng 1999; 87:307-314.
44. Minko I, Holloway SP, Nikaido S et al. Renilla luciferase as a vital reporter for chloroplast gene expression in Chlamydomonas. Mol Gen Genet 1999; 262:421-425.

CHAPTER 5

Influence of Codon Bias on the Expression of Foreign Genes in Microalgae

Markus Heitzer,* Almut Eckert, Markus Fuhrmann and Christoph Griesbeck

Abstract

The expression of functional proteins in heterologous hosts is a core technique of modern biotechnology. The transfer to a suitable expression system is not always achieved easily because of several reasons: genes from different origins might contain codons that are rarely used in the desired host or even bear noncanonical codons, or the genes might hide expression-limiting regulatory elements within their coding sequence. These problems can also be observed when introducing foreign genes into genomes of microalgae as described in a growing number of detailed studies on transgene expression in these organisms. Particularly important for the use of algae as photosynthetic cell factories is a fundamental understanding of the influence of a foreign gene's codon composition on its expression efficiency. Therefore, the effect of codon usage of a chimeric protein on expression frequency and product accumulation in the green alga *Chlamydomonas reinhardtii* was analyzed. This fusion protein combines a constant region encoding the zeocin binding protein Ble with two different gene variants for the green fluorescent protein (GFP). It is shown that codon bias significantly affects the expression, but barely influences the final protein accumulation in this case.

General Aspects of Codon Bias in Pro- and Eukaryotic Expression Hosts

The genetic code is degenerated, i.e., there are multiple codons for the same amino acid. This allows organisms to select a subset of codons to efficiently encode all their proteins and to leave other base triplets underrepresented. Moreover, within one single organism the codon bias can vary between abundantly and moderately expressed genes and between long and short genes. For *E. coli*, the far most used prokaryotic expression host, the codon usage observed especially for Gly, Arg and Pro in the mRNA population of highly abundant genes is closely reflected by the corresponding tRNA populations.[1] Such a severe codon bias has been discussed for a long time as one possible hindrance in the high level expression of heterologous genes in general. It was reasoned that insufficient tRNA pools can lead to translational stalling, premature translation termination, translation frame shifting and amino acid misincorporation.[2] For example, multiple consecutive rare codons near the N-terminus of a coding sequence led to significant reduction in the expression efficiency of this gene.[3] In another case, supplementing the endogenous tRNA population of *E. coli* by additional plasmid encoded copies caused a strong increase in the expression yield of human tissue plasminogen activator.[4]

*Corresponding Author: Markus Heitzer—Universität Regensburg, Kompetenzzentrum für Fluoreszente Bioanalytik, Josef - Engert - Str. 9, 93053 Regensburg, Germany.
Email: markus.heitzer@exfor.uni-regensburg.de

Transgenic Microalgae as Green Cell Factories, edited by Rosa León, Aurora Galván and Emilio Fernández.

Effects of codon bias on the efficiency of endogenous and foreign gene expression are not limited to prokaryotic hosts, but have also been observed in different yeast expression systems,[5-7] higher plants like maize[8] or *Arabidopsis*,[9] animals like *Drosophila*[10] and mammalian cell lines.[11]

Although there have been several reports about the introduction of foreign genes into the genomes of various microalgae, stable genetic transformation of microalgal genomes is achieved routinely only for very few species. The use of microalgae for the efficient production of valuable compounds like recombinant proteins or their metabolic engineering using synthetic enzyme variants will depend largely on the efficiency of foreign gene expression. This efficiency is likely to be influenced by the codon composition of the transgene, as it is the case for many other organisms. Clearly, the detailed analysis of codon usage on expression has always to rely on wide information about the genome and even more the frequency of individual codons in abundant and average mRNAs. Yet, this combination of knowledge is available only for the diatom *Phaeodactylum tricornutum* and the green alga *Chlamydomonas reinhardtii*. For some other species genetic transformation has been reported, but the necessary information about codon frequency in mRNAs is restricted. Moreover, for some algae the number of experiments on transgene expression is still limited, or only transient expression has been observed (see also Chapter 1). Therefore judging the influence on expression efficiency of transgenes for *Cyclotella, Navicula, Cylindrotheca* or *Thalassiosira* species is still highly speculative.

Phaeodactylum tricornutum

The first reports about transformation of the *P. tricornutum* nuclear genome with the bacterial phleomycin resistance gene *ble* and the reporter chloramphenicol acetyltransferase demonstrated a stable integration into the nuclear genome.[12] These findings suggested a continuous expression of the foreign genes without the need for constant selection regardless of any differences in codon usage between donor and host organism. In contrast, similar experiments indicated unstable expression of the bacterial *S. hindustanus ble* gene in about 80% of all lines.[13] These effects are likely to be explained by one or more different epigenetic silencing mechanisms, a common phenomenon in transgene expression in plants and algae. A general inactivation of foreign genes could be excluded by the analysis of a second reporter gene, firefly luciferase. This gene was introduced in parallel and did not show a corresponding silencing effect. Both genes again originate from heterologous sources and show a different codon usage. The bacterial *ble* presents a very high GC-content (~70%), whereas the luciferase gene from *Photinus* has a much lower GC-content (~37%). Surprisingly, the average codon usage for *Phaeodactylum* as calculated from the Kazusa codon usage data base (http://www.kazusa.or.jp/codon/) is intermediate (54%), but with a preference for cytosine in the third position particularly in abundant transcripts.[14] In further reports, the expression of several GFP genes and glucose uptake proteins from different organisms were analyzed in detail. Wildtype *gfp* from *Aequorea* and a codon modified variant for maximal expression in *Arabidopsis* (*mgfp4*) did not result in detectable protein, but visible amounts of GFP were obtained by using *egfp*, a codon optimized form for human cells. Since contrary to *gfp* and *mgfp4* the codon composition of *egfp* is very close to that of *P. tricornutum*, the successful expression of *egfp* could be directly attributed to its codon bias.[15] In a second approach, the authors found that yeast glucose transporters (Hxt1, Hxt2 and Hxt4, ~39% GC) were not expressed, but the human erythrocyte Glut1 (~59% GC) and the *Chlorella* Hup1 (~61% GC) hexose transporters were produced efficiently, allowing heterotrophic growth of *P. tricornutum* on glucose as the only carbon source. The authors speculated, that these results "may reflect differences in codon usage between yeast and *P. tricornutum*".[16]

Chlamydomonas reinhardtii —Expression from Chloroplast and Nucleus

One microalga that has attracted attention as a promising cell factory system is *Chlamydomonas reinhardtii*, especially in view of the broad genetic and molecular toolset available for this organism. However, it has often been speculated that in this species a strong bias for specific codons in highly expressed nuclear and chloroplastic genes may significantly influence the expression efficiency of foreign genes after integration into any of the genomes. For example, only genes with a high GC-content similar to that found in *C. reinhardtii* (GC~61%) and with a low percentage of rarely used codons could be established as nuclear selection markers.[17]

Some systematic studies have been carried out for genes expressed from the chloroplast genome. It was demonstrated that a variant of GFP adapted to the chloroplast codon usage accumulated about 80-fold more in chloroplasts of transformed strains as compared to the non adapted gene.[18] The integration into the chloroplast genome occurs via homologous recombination, and therefore positional effects of genome integration can be excluded. Further experiments using different genes (for bacterial luciferase[19] and a full length single chain antibody)[20,21] optimized for the chloroplast codon usage were comparably successful. These findings suggest a significant influence of codon composition on the expression efficiency of a transgene when incorporated into the chloroplast genome of *Chlamydomonas* (see also Chapters 4 and 8).

Comparable experiments have been performed for nuclear encoded genes in *Chlamydomonas*. A first report using a synthetic gene encoding the green fluorescent protein from *Aequorea victoria* indicated a positive effect of the adaptation of codons to the nuclear bias of the alga.[22] In this study, no comparison with the AT-rich original gene was possible, because expression of the unmodified gene failed under the same promoter. Equal results were obtained for GFP expression in the closely related multicellular green alga *Volvox carteri* (GC~60%), where only the codon optimized gene variant produced a detectable amount of protein.[23] The successful use of synthetic genes with improved codon usage was continued in further experiments using a synthetic gene for the luciferase of *Renilla reniformis* adapted to the nuclear codon frequency of *C. reinhardtii*.[24]

In order to study the influence of codon usage on foreign gene expression from the nuclear genome in more detail, we compared the production of two variants of the GFP, the original AT-rich variant (*mgfp*, 39% GC) of the *Aequorea* gene and the synthetic *cgfp* (62% GC) with *Chlamydomonas* adapted codon usage (Fig. 1).[22] The *mgfp* was created from the plasmid pGFP (BD Clontech, USA) by introducing three individual mutations (F64L, S65T and T203I), which were also present in *cgfp*. As expression of *mgfp* failed before, both genes were cloned in frame with the bacterial *ble* gene from *S. hindustanus* in the previously described vector pSP124S-M,[22] allowing the direct selection of zeocin resistant clones expressing the Ble-GFP fusion proteins from the same constitutive promoter. Four independent transformations of *Chlamydomonas cw15 arg⁻* were carried out as described[24] with the two gene variants transformed in parallel using the same culture divided into equal portions for comparison. Selection of zeocin resistant colonies was performed after an 18 h recovery phase in liquid nonselective medium on standard TAP (Tris-Acetate-Phosphate) medium supplemented with arginine (90 µg/ml) and zeocin (5-20 µg/ml) for about 10-14 days under constant illumination at 25°C.

The number of zeocin resistant clones was determined after two and four weeks (Table 1). This number was dependent on the variant of *gfp* that had been used for transformation, as the codon adapted gene resulted in a nearly 5-fold elevated number of resistant colonies. Obviously, there was a difference between the two lines of transformants, as determined by their growth after transfer from nonselective to selective conditions: whereas most of the cGFP-transformants showed very good growth on 5 µg/ml of zeocin (23 of 27 lines) none of the mGFP-strains grew as well. In this experiment 5 µl of liquid culture pregrown in nonselective medium for one week was spotted onto a selective plate and growth was monitored after five days. All three mGFP lines showed only sparse spotty growth, where most cGFP lines grew to dense green colonies (Table 2). This growth difference indicated a lower expression of

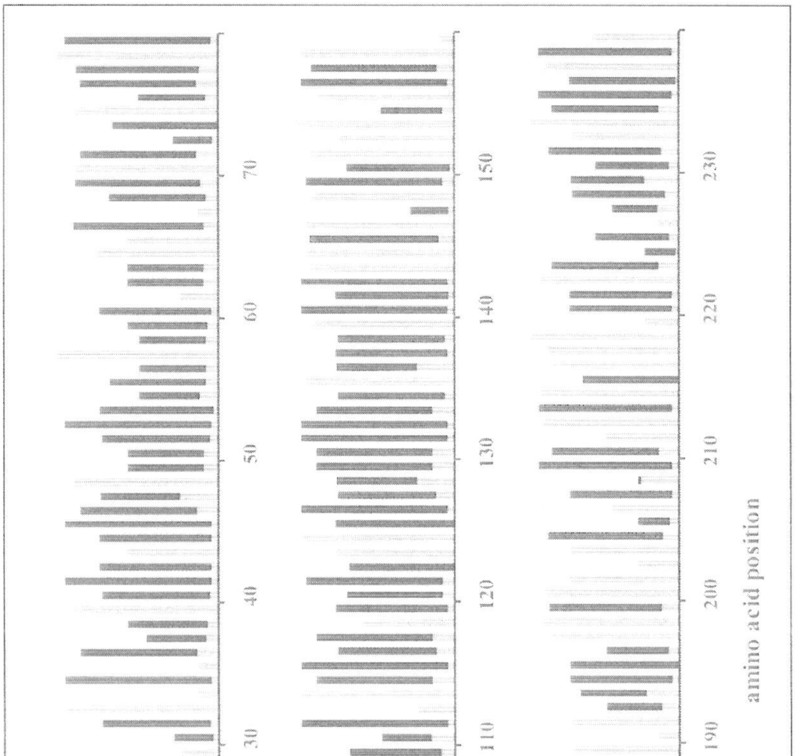

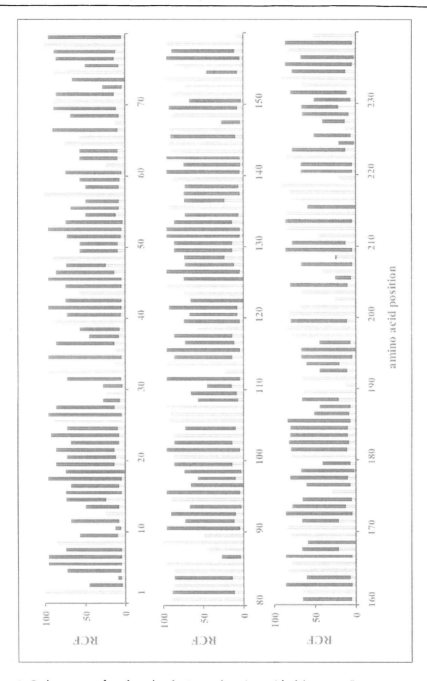

Figure 1. Codon usage of *mgfp* and *cgfp*. For each amino acid of the green fluorescent protein the relative frequency of individual codons (RCF) in the nuclear genome of *Chlamydomonas reinhardtii* is displayed. The *mgfp* sequence from *Aequorea* (in gray) contains several stretches with rare codons, which were replaced in the synthetic *cgfp* (in black) by adaptation of the gene to the nuclear codon usage of *Chlamydomonas*. The graphical codon usage analyser (gcua) tool was used for calculation.[24]

Table 1. Total numbers of zeocin resistant* Chlamydomonas reinhardtii *transformants obtained with two variants of* ble-gfp *from four independent transformations

To compare the transformation efficiency from different experiments, a comparison value (CV) was defined as

$$CV = \frac{number\ of\ clones\ (N)}{number\ of\ cells\ (n) \cdot amount\ on\ plasmid\ (p)} \left[\frac{1}{10^{-7} \cdot \mu g^{-1}}\right]$$

For each experiment 1-4·10^7 cells (n) were transformed with 0.6-2.5 μg plasmid (p).

Plasmid	N (Two Weeks)	CV	Expressing Clones (Further Cultivated)	N (Four Weeks)
pSP124S-mGFP	27	30	3	2
pMF124-cGFP	105	160	27	26

Table 2. Growth assay of primary transformants after a period of nonselective growth

Plasmid / Growth	+++	++	+	-
pSP124S-mGFP	0	0	2	1
pMF124-cGFP	20	3	3	1

Cells pregrown in liquid culture under nonselective conditions were transferred onto a selective plate. Growth was monitored after a period of five days in continuous light at 25°C and was grouped into four categories: +++ very dense growth, absolutely round shaped colony; ++ dense growth, not completely filled colony; + sparse growth, individual spots visible; - no growth detectable.

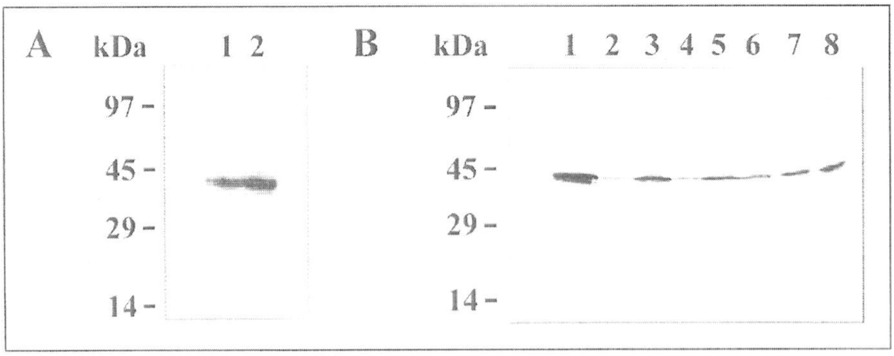

Figure 2. Western blot analysis of Ble-mGFP and Ble-cGFP lines with an anti-Ble-antibody. Samples with the same amount of protein according to a culture volume of approx. 30 μl were analyzed. A. Ble-mGFP lines. Lanes 1 and 2 correspond to clones AA1 and AA2. B. Ble-cGFP lines. Lanes 1-8 correspond to clones AA11 and AA15-21.

Ble-mGFP after a period of nonselective growth. Possible reasons for this observation could be active transcriptional or post-transcriptional gene silencing, but also stalled or prematurely terminated translation of the Ble-mGFP.

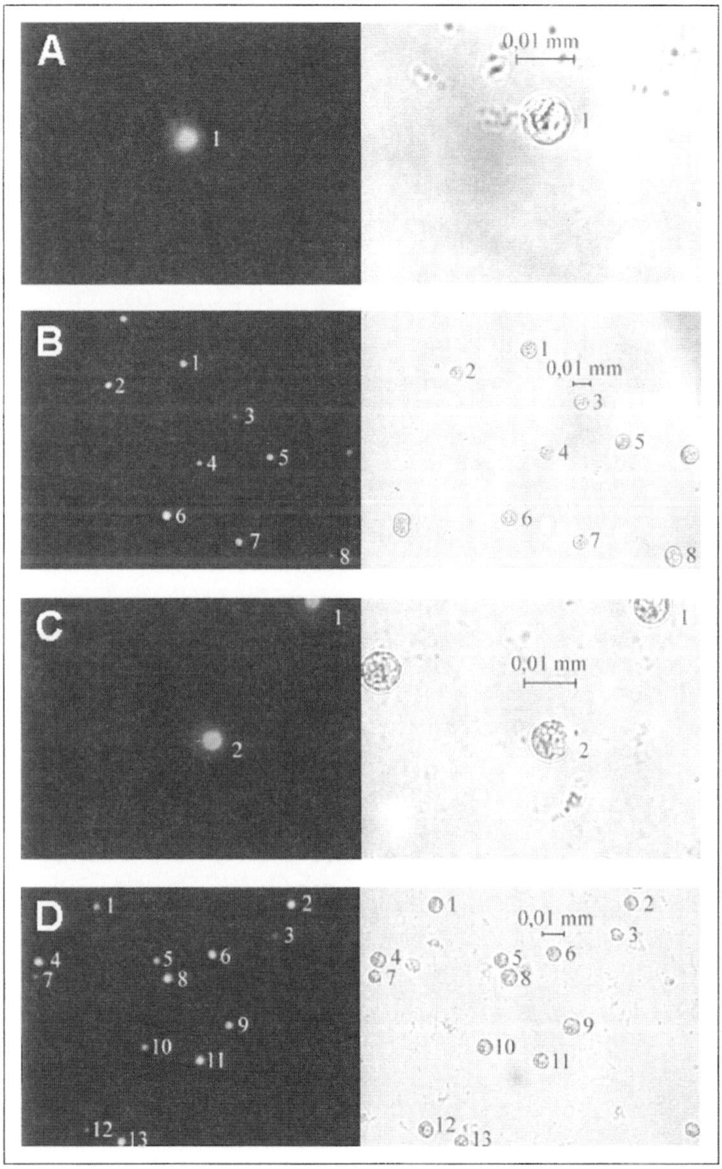

Figure 3. Fluorescence microscopy of Ble-mGFP and Ble-cGFP lines. A) and (B) Ble-mGFP clone AA2. The photographs show the localization of Ble-mGFP in the nucleus of a single cell (A) and with lower magnification expression levels in different cells (B) by fluorescence (left side) and light microscopy (right side). Identical cells are marked with the same numbers. C and D. Equal settings for Ble-cGFP clone AA11. There is no obvious difference detectable between these two strains. For detailed experimental procedures see reference 22.

To further analyze the expression of the Ble-GFP fusion protein both immunoblot assay and fluorescence microscopy were performed. Protein samples of all 28 lines grown in selective medium were separated by SDS-PAGE, transferred to nitrocellulose membranes and probed with a primary anti-Ble antibody (Cayla, France; dilution 1:1000). Detection of bound antibody was performed after three washing steps using an alkaline phosphatase conjugated secondary anti-rabbit-IgG antibody (Sigma, USA, dilution 1:1000) and a color precipitation assay. The results are shown in Figure 2. The Ble-GFP fusion protein was clearly detected at the correct size (43 kDa) for both resistant strains. No indication for premature termination of translation or protein degradation resulting in different products of various sizes could be observed. Also no significant difference between *mgfp* and *cgfp* lines in the amount of expressed protein as estimated from signal intensity was determined, although the expression levels of the *cgfp* clones covered a broad range.

The observed growth differences on selective medium could also result from a wide variation of expression efficiency within a single line, i.e., some cells showing a high expression level, some a very low. The Ble-GFP fusion protein can be directly visualized using fluorescence microscopy of living cells. We analyzed multiple sets of cells from different transformants of both gene variants and compared the expression levels (Fig. 3). No significant difference between the two types of transformants was obvious; all observed cells showed a comparably high level of GFP as determined by intensity measurement of fluorescence in the cell nucleus. However, an exact quantification of the three-dimensional fluorescence intensity and the conclusion for the expression level in the cell could not be achieved with this method.

Concluding Remarks

As observed in many other organisms, there are indications for a strong effect of codon usage on the expression efficiency of transgenes in *Chlamydomonas reinhardtii* for both nuclear and chloroplast encoded transgenes. Since the number of selectable transformants is significantly higher for codon optimized nuclear genes, the codon usage is likely to have an effect on the decision if a transgene is expressed or not and maybe also on the decision if silencing of this transgene takes place or not. So far, not adapted genes seem to be more susceptible to silencing effects. However, in clones that express transgenes, no differences regarding the expression quantity and quality can be observed between codon optimized and unmodified genes. But studies of the influence of codon bias on nuclear expression are always complicated by positional effects, since expression of transgenes strongly depends on the position within the genome and integration of foreign genes into the chromosomes takes place randomly in *C. reinhardtii*. Hence, a definite statement about codon effects cannot be made. Targeted integration of transgenes into the genome by homologous or Cre/*lox*-mediated site-specific recombination could be a helpful tool to circumvent this problem.[25]

References

1. Dong H, Nilsson L, Kurland CG. Covariation of tRNA abundance and codon usage in Escherichia coli at different growth rates. J Mol Biol 1996; 260(5):649-63.
2. Kurland C, Gallant J. Errors of heterologous protein expression. Curr Opin Biotechnol 1996; 7(5):489-93.
3. Goldman E, Rosenberg AH, Zubay G et al. Consecutive low-usage leucine codons block translation only when near the 5' end of a message in Escherichia coli. J Mol Biol 1995; 245(5):467-73.
4. Novy R, Drott D, Yeager K. Overcoming the codon bias of E. coli for enhanced protein expression. Innovations 2001; 12:1-3.
5. Sinclair G, Choy FY. Synonymous codon usage bias and the expression of human glucocerebrosidase in the methylotrophic yeast, Pichia pastoris. Protein Expr Purif 2002; 26(1):96-105.
6. Janatova I, Costaglioli P, Wesche J et al. Development of a reporter system for the yeast Schwanniomyces occidentalis: Influence of DNA composition and codon usage. Yeast 2003; 20(8):687-701.
7. Friberg M, von Rohr P, Gonnet G. Limitations of codon adaptation index and other coding DNA-based features for prediction of protein expression in Saccharomyces cerevisiae. Yeast 2004; 21(13):1083-93.

8. Viotti A, Balducci C, Weil JH. Adaptation of the tRNA population of maize endosperm for zein synthesis. Biochim Biophys Acta 1978; 517(1):125-32.
9. Chiapello H, Lisacek F, Caboche M et al. Codon usage and gene function are related in sequences of Arabidopsis thaliana. Gene 1998; 209(1-2):GC1-GC38.
10. Carlini DB, Stephan W. In vivo introduction of unpreferred synonymous codons into the Drosophila Adh gene results in reduced levels of ADH protein. Genetics 2003; 163(1):239-43.
11. Slimko EM, Lester HA. Codon optimization of Caenorhabditis elegans GluCl ion channel genes for mammalian cells dramatically improves expression levels. J Neurosci Methods 2003; 124(1):75-81.
12. Apt KE, Kroth-Pancic PG, Grossman AR. Stable nuclear transformation of the diatom Phaeodactylum tricornutum. Mol Gen Genet 1996; 252(5):572-9.
13. Falciatore A, Casotti R, Leblanc C et al. Transformation of nonselectable reporter genes in marine diatoms. Mar Biotechnol (NY) 1999; 1(3):239-251.
14. Montsant A, Jabbari K, Maheswari U et al. Comparative genomics of the pennate diatom Phaeodactylum tricornutum. Plant Physiol 2005; 137(2):500-13.
15. Zaslavskaia LA, Lippmeier JC, Kroth PG et al. Transformation of the diatom Phaeodactylum tricornutum (Bacillariophyceae) with a variety of selectable marker and reporter genes. Journal of Appl Phycol 2000; 36(2):379-386.
16. Zaslavskaia LA, Lippmeier JC, Shih C et al. Trophic conversion of an obligate photoautotrophic organism through metabolic engineering. Science 2001; 292(5524):2073-2075.
17. Leon-Banares R, Gonzalez-Ballester D, Galvan A et al. Transgenic microalgae as green cell-factories. Trends Biotechnol 2004; 22(1):45-52.
18. Franklin S, Ngo B, Efuet E et al. Development of a GFP reporter gene for Chlamydomonas reinhardtii chloroplast. Plant J 2002; 30(6):733-44.
19. Mayfield SP, Schultz J. Development of a luciferase reporter gene, luxCt, for Chlamydomonas reinhardtii chloroplast. Plant J 2004; 37(3):449-58.
20. Mayfield SP, Franklin SE, Lerner RA. Expression and assembly of a fully active antibody in algae. Proc Natl Acad Sci USA 2003; 100(2):438-42.
21. Mayfield SP, Franklin SE. Expression of human antibodies in eukaryotic micro-algae. Vaccine 2005; 23(15):1828-32.
22. Fuhrmann M, Oertel W, Hegemann P. A synthetic gene coding for the green fluorescent protein (GFP) is a versatile reporter in Chlamydomonas reinhardtii. Plant J 1999; 19(3):353-61.
23. Ender F, Godl K, Wenzl S et al. Evidence for autocatalytic cross-linking of hydroxyproline-rich glycoproteins during extracellular matrix assembly in Volvox. Plant Cell 2002; 14(5):1147-60.
24. Fuhrmann M, Hausherr A, Ferbitz L et al. Monitoring dynamic expression of nuclear genes in Chlamydomonas reinhardtii by using a synthetic luciferase reporter gene. Plant Mol Biol 2004; 55(6):869-81.
25. Zorin B, Hegemann P, Sizova I. Nuclear-gene targeting by using single-stranded DNA avoids illegitimate DNA integration in Chlamydomonas reinhardtii. Eukaryot Cell 2005; 4(7):1264-72.

CHAPTER 6

In the Grip of Algal Genomics

Arthur R. Grossman*

Abstract

Algae are dominant primary producers on the Earth and have a major impact on global productivity and biogeochemical cycling. There are still few algal genomes that have been completely characterized, and resources directed toward algal genomic sequencing are limited. However, it is also becoming evident that algae and prokaryotic picoplankton have a critical role in the fixation and sequestration of carbon, and so the interest in algal genomics is expanding. There are some algae for which full or near-full genome sequences have been secured; these genomes include those of the red alga *Cyanidioschyzon merolae*, the green algae or chlorophytes *Chlamydomonas reinhardtii* and *Volvox carteri*, the marine picoeukaryote *Ostreococcus tauri* (two different strains of *O. tauri* have been sequenced), the diatoms *Thalassiosira pseudonana* and *Phaeodactylum tricornutum*, and the haptophyte *Emiliania huxleyi*. There is also a full sequence for the vestigal 'red' algal genome of the nucleomorph of the Cyptomonad *Guillardia theta*. In addition, numerous genomes of photosynthetic microbes, including marine *Synechococcus* and *Prochlorococcus* species have been sequenced. There have also been projects developed to define algal transcriptomes as determined by cDNA analysis, full genome sequences of numerous plastids, and the genomes of a variety of viruses that infect marine and freshwater algae. The recent efforts focused on acquiring and analyzing algal genome sequences have generated an influx of exciting data to a field that is in its infancy. In this review I discuss potential criteria for determining which organisms should be targeted for genome projects, successful forays into algal genomic sequencing, and some of the inferences generated from the analysis of the sequence information.

Introduction

Genomics involves defining sequences of genes and genomes through the analysis of genomic or cDNAs libraries, and then analyzing that information to learn about the architecture, organization, evolution, expression and function of genes and genomes. Such information provokes formulations of hypotheses concerning ways in which organisms have adapted to particular physiological niches, relationships among diverse groups of organisms, and the molecular events that accompany developmental processes. The accumulation of genomic and cDNA information has made it possible to monitor genome-wide expression, using high density cDNA or oligonucleotide-based microarrays. While several model eukaryotic and prokaryotic systems have been extensively characterized at the genomic level, the major genomic efforts have been directed toward sequence analysis of the human genome and genomes of human pathogens. The development of better methods for high-throughput sequencing of large DNA fragments and the reduced cost of sequencing reactions have allowed full genome sequence analysis for a

*Arthur R. Grossman—The Carnegie Institution, Department of Plant Biology, 260 Panama Street, Stanford, CA 94305, U.S.A. Email: arthurg@stanford.edu

Transgenic Microalgae as Green Cell Factories, edited by Rosa León, Aurora Galván and Emilio Fernández.

diverse group of organisms, including those of plants, nonpathogenic bacteria and algae. Indeed, more researchers are becoming aware of the numerous contributions made by algae to terrestrial, aquatic and atmospheric habitats of the Earth. We are beginning to understand how important these organisms are as dominant primary producers and the critical role that they play in biogeochemical cycling of resources.

The algae include a group of photosynthetic organisms present in a range of aquatic and terrestrial environments. These morphologically varied organisms can be vanishingly small (tiny picoplankton that inhabit open oceans)[1,2] (also see http://www.sb-roscoff.fr/Phyto/PICODIV/PICODIV_publications.html)] or towering plant-like structures that form forests in coastal waters.[3] Indeed, diversity within this group of organisms is enormous. The algae include individuals at extremes with respect to size and shape, and they house of variety of novel metabolic pathways and synthesize various chemical compounds of both ecological and commercial importance. One striking example of this diversity concerns the range of pigment molecules associated with light-harvesting antenna complexes that drive photosynthetic electron transport. In the green algae, light is harvested for photosynthesis by antenna complexes that mostly contain the pigments chlorophylls a and b and various carotenoids (β-carotene, lutein, xanthophylls), while the antennae pigments of red algae and cyanobacteria are predominantly phycobiliproteins; these proteins are covalently bound to the bilin chromophores phycerythrobilin, phycocyanobilin or phycourobilin. In contrast, dominant and diverse xanthophyll species (oxygenated carotenoids) are major constituents of light-harvesting antenna complexes in diatoms and dinoflagellates. In addition, the cell wall structure and composition, along with the characteristics of stored polysaccharides and lipids exhibit strong diversity among algae; some algae having cell walls that are predominantly cellulosic while others have proteinaceous or silicacious walls.

In many parts of the world algae are processed and used as a food source. They may be included in salads and soups, but are most well known as being the wrap for sushi (nori), which is made from fronds of the red alga *Porphyra.* Algae are rich in vitamins, carotenoids and antioxidants, and are commonly processed and marketed by the health food industry (http://www.1001beautysecrets.com/nutrition/algae/)(http://www.crystalpurewater.com/health.htm). Algae are also used as additives to feed for aquaculture, providing an appealing pigmentation that permeates the flesh and skin of fish. In the medical sciences, the phycobiliproteins have been used to confer fluorescent properties to tags for localizing and quantifying specific surface antigens. The polysaccharides that accumulate in some algae include such compounds as agar, alginates and fucoids.[4-7] These polysaccharides have been used as anticoagulants,[7] to make gel matrices for the delivery of medicines, for preparing solid medium to maintain and grow bacteria, as thickeners added to ice cream and as the base for cosmetics. Algal products are also added to cleaners, ceramics and toothpaste (http://www.nmnh.si.edu/botany/projects/algae/Alg-Prod.htm). Furthermore, diatoms and dinoflagellates synthesize long chain polyunsaturated fatty acids ('fish oils') that have been incorporated into baby formula in many countries throughout the world as they appear beneficial for mammalian brain development.[8,9]

Algae also play a role in the stabilization of various ecosystems. They participate in symbiotic associations, feeding heterotrophic host organisms various forms of fixed carbon. The dinoflagellate *Symbiodinium,* which lives within the tissue of coral polyps, fixes inorganic carbon and transfers much of the fixed carbon to the coral host. The feeding of the animal by the intracellular endosymbiont is essential for establishing and maintaining the coral reef community, which can physically stabilize coastal environments.[10] Rising biosphere and oceanic temperatures are causing degradation of many biotic systems including the algal symbiotic associations that sustain coral populations (the corals 'bleach'); the loss of coral reefs is having a pronounced and immediate impact on coastal environments.[11] There is also considerable concern about controlling the growth of specific algae in the oceans and lakes since they can attain high densities and form blooms that stimulate the proliferation of consumers and the generation of anoxic conditions that suffocate aquatic animals. Furthermore, groups of algae and cyanobacteria produce neurotoxins that are threatening global water supplies. In the long term, the changing climate of terrestrial and oceanic environments will drive significant changes in phytoplankton populations, which in turn will have a profound impact on global carbon fluxes and the trophic transfer of carbon in food chains.

Which Organisms Should Have Their Genomes Sequenced?

Genomics has advanced from full-length sequencing of individual genes and clones, to the sequencing of whole genomes and now, with the development of metagenomic approaches, to sequencing populations of DNA fragments derived from environmental samples. Random sequencing of environmental DNA samples can be valuable for gene discovery and for unveiling biological processes potentially important for survival in specific ecological niches.[12-14] However, it is still full-genome sequence information that provides the blueprint for developing a broad vision of the genetic potential of a specific organism.

Whether an organism should be considered for genomic studies will depend upon many issues, including its ecological, evolutionary and economic importance, whether or not it can be genetically manipulated, how easy it is to culture, and how much this organism has already been studied. A number of factors to be considered when deciding on organisms to be chosen for full genome sequencing are given in Table 1.

Full Genome Sequences

Full genome sequence and cDNA information are being generated by consortia of researchers, often in conjunction with large sequencing facilities. There has been a rapid growth of databases containing both algal plastids and nuclear sequences. The sequence of many plastid genomes can be readily obtained, including those of the green algae *Chlamydomonas reinhardtii* (http://www.chlamy.org/chloro.html), *Nephroselmis olivacea,*[17] *Chaetosphaeridium globosum,*[18] *Chlorella vulgaris*[19] and *Mesostigma viride,*[20] the cryptophyte *Guillardia theta,*[21] the stramenopile (diatom) *Odontella sinensis,*[22,23] the glaucophyte *Cyanophora paradoxa,*[24] the red alga *Cyanidium caldarium*[25] and the euglenophyte *Euglena gracilis.*[26] The plastid genes of dinoflagellates are unique in that each gene appears to reside on its own minicircle.[27,28] The plastid (and mitochondrial) genome sequences of algae and plants can be accessed at http://megasun.bch.umontreal.ca/ogmp/projects/other/cp_list.html.

There is much less information concerning the nuclear genomes of algae, although knowledge in this area is rapidly expanding. Genomes for which there is a complete or near complete sequence, that is readily accessible to the public, include those of the red alga *Cyanidioschyzon merolae* (http://merolae.biol.s.u-tokyo.ac.jp/),[29] the diatom *Thalassiosira pseudonana* (http://genome.jgi-psf.org/thaps1/thaps1.home.html),[30] and the green alga *Chlamydomonas reinhardtii* (http://genome.jgi-psf.org/Chlre3/Chlre3.home.html). Other genomes either sequenced and not yet released or in the process of being sequenced include those of *Ostreococcus tauri* (http://www.iscb.org/ismb2004>/posters/stromATpsb. ugent.be_844.html), *Volvox carteri, Emiliania huxleyi, Phaeodactylum tricornutum* (see http://trace.ensembl.org/perl/traceview?attr=tt_ce_sp&tt_1=1), *Aureococcus anophagefferens* and *Micromonas pusilla* (http://www.microbialgenome.org/organisms.shtml). In additions, the nucleomorph of *Guillardia theta,* which represents a vestigial red algal genome, has three chromosomes that have been sequenced.[21] But the sequences completed to this point represent the initial stages in an explosion of sequence information that is and will continue to embrace organisms within the different kingdoms of life. Some of the algae for which extensive genome sequence information is available are discussed below.

Chlamydomonas reinhardtii

Genetic, molecular, physiological and genomic features have made *Chlamydomonas reinhardtii*, a unicellular, green alga in the family Volvocales, a popular model system that has been developed over the last half century. This organism has been exemplary for the study of

Table 1. Deciding on algal genomes for sequencing

Property	Comments
1. Culturing the organims	Many algae are not easy to grow in the laboratory, including many of the large macrophytic algae. It is advantageous to be able to grow the organisms on defined medium, which would be useful when examining metabolism and acclimation to environmental conditions.
2. Control of sexual reproduction	Many marine algae have complex life histories and often the sexual cycle is difficult to control. The ability to control sexuality would afford investigators the opportunity to engineer crosses, which would help unmask recessive mutations in diploid organisms and facilitate map-based cloning of specific mutant alleles.
3. Mutant generation	The generation and characterization of mutants often helps to define specific biological processes. Genetic analyses of strains harboring mutations of interest would allow researchers to identify the gene modified in the mutant organism and the nature of that modification.
4. Uninucleate versus coenocytic	It is likely to be easier to genetically manipulate algal forms species that are uninucleate, although some of the coenocytic forms are extremely interesting with respect to wound healing responses and the arrangement and movement of organelles within giant cells.
5. Prior knowledge	Genomic information may be better exploited if there is already a knowledge-base with respect to the ecology, physiology and genetics of the organism. Prior knowledge may provide information about physiological processes and the ecological milieu in which the organism thrives, which in turn can help researchers interpret the functions of genes and potential metabolic and regulatory pathways.
6. Evolutionary aspects	It would be interesting to investigate genera comprised of different species (e.g., species in which there are many different ecotypes), which could lead to strong comparative genomic analysis and insights into the evolution of species/ecotypes and provide critical information relating to niche specialization. It would also be valuable to have full genome sequences for organisms positioned at evolutionary branchpoints, as well as those for which there is an extensive fossil record; the latter could be used to calibrate the evolution of specific biological features (when in evolutionary time they occurred) that characterize a particular genus.
7. Ecological importance	Concentrating genomic studies on the dominant organisms in aquatic and terrestrial ecosystems would provide a better understanding of the physiology of these organisms and the ways that they potentially interact with other organisms in their environment. This knowledge would also help in designing strategies to manage the health of specific ecosystems.
8. Economic importance	Valuable products are derived from algae, and in many parts of the world they serve as a source of nutrition. Genomic information about these organisms may allow for tailoring nutritional composition and lead to the biosynthesis of new and valuable compounds (polysaccharides, carotenoids, polyunsaturated fatty acids).

Table continued on next page

Table 1. Continued

Property	Comments
9. Specific attributes	Some organisms have attributes that may be of intrinsic biological or of potential commercial interest; such attributes might include the patterning of cell walls in diatoms, the production of long chain fatty acids in dinoflagellates and the synthesis and secretion of coccoliths.
10. Genome size and repeat structure	There is a strong bias toward sequencing genomes that aren't too large. This reduces both the cost of the project and the difficulties associated with the assembly of the sequence data. For organisms with large genomes, developing a cDNA sequencing project, or enriching for expressed regions of the genome[15,16] might be appropriate.
11. Establish strong infrastructure	Leadership and infrastructure are critical to the success of a genome project. The community, which often has little experience with the technologies associated with acquiring and analyzing genomic information, must generate links to sequencing centers and recruit experts to organize and mine the data.

various key biological processes associated with animals and plants; this is especially the case in the areas of photosynthesis and flagella biogenesis and function. *C. reinhardtii* has had the advantage of being used as a model genetic system starting from the mid 20th century, and is now also amenable to applications of both molecular and genomic technologies. Plastid and nuclear genomes of this alga can be transformed by various procedures,[31-35] and several markers are available for the selection of transformants[31,36-43] (see also Chapters 1 and 4). Numerous recombinant libraries in plasmids, cosmids, and BACs have been generated and analyzed,[44-47] and can be used to rescue the phenotypes of specific *C. reinhardtii*[48-50] or *Escherichia coli*[51,52] mutants. Random insertion of specific marker genes can be used to generate mutants in which the mutant alleles are tagged by the introduced DNA[53-62] (see also Chapter 7), and alleles not tagged that result from point mutations or small indels can be identified by map-based cloning.[63,64] Levels of specific transcripts can be depressed using antisense or RNAi suppression strategies,[65-68] and a number of reporter genes are available that can help elucidate promoter function and mechanisms that underlie gene regulation.[69-79] Integration of genes into the chloroplast genome[80,81] occurs by homologous recombination, which has enabled researchers to target specific genes for inactivation and to introduce specific lesions into genes for evaluating gene function.[82-94] One of the most extensive uses of *C. reinhardtii* has been in the dissection of photosynthetic processes. The features of this organism most advantages for the analysis of photosynthesis are: (1) it can grow heterotrophically in the dark using acetate as a sole fixed carbon source, (2) dark-grown cells maintain normal chloroplast structure and function, (3) mutations that adversely affect photosynthetic function are readily isolated and characterized[95,96] and (4) in vivo spectroscopy can be used to help define specific aspects of photosynthetic function in both wild-type and mutant strains. Technique for genetic manipulation of *C. reinhardtii*, first developed by Sager,[97] were used by Levine and colleagues to elucidate photosynthetic electron transport and CO_2 fixation using mutants defective for photosynthesis.[98-105] Many mutants defective for cell motility were found to have defects in specific components of the flagella, a finding that has helped delineate structural and functional features of the flagella. Exciting new work has demonstrated a relationship between specific flagella polypeptides and mammalian diseases; these polypeptides are associated with cilia in mammalian cells.[106-109]

Over the last decade the development of DNA microarray or chip technology has allowed the analysis of global gene expression. The generation of cDNA and genomic information for *C. reinhardtii*[47,110-114] (http://genome.jgi-psf.org/chlre1) has faciliated DNA microarray and macroarray construction, which has already been used to study various biological processes.[115-121] Genome-wide and proteomic approaches are being used to understand the dynamics of the photosynthetic apparatus in response to nutrient conditions,[115,116,122] light and circadian programs,[115,123-125] composition of pigment protein complexes,[126,127] identification of components involved in iron assimilation[128] and the polypeptide constituents of the flagella and basal body.[109]

There is a wealth of information contained within the genomic sequence of *C. reinhardtii*. The genome is approximately 120 Mb, and although the sequence still has numerous gaps, most of the gaps should be filled in by the time this review appears. The genome contains approximately 16,000 gene models, most supported by extensive EST data, with a low level of nuclear plastid DNA segments or NUPTs, relative to the *A. thaliana* or *O. sativa* genomes.[129] *C. reinhardtii* genes often have long 3' untranslated sequences and numerous introns. Both cDNA and genomic sequence information has helped elucidate genes encoding both LHCB and LHCA polypeptides (for photosystem II and photosystem I, respectively).[127] Genes encoding ELIPs and other polypeptides thought to be associated with light-harvesting complexes, are likely critical for the management of absorbed excitation energy; these include LI818 or LHCSrc,[127,130,131] and PSBS,[132] a four transmembrane member of the *LHC* gene family involved in xanthophyll cycle-dependent quenching.[133-135] Numerous genes encoding proteins critical for chromatin structure, nutrient acquisition (nitrogen, sulfur, phosphorus, iron) and assimilation, and carbon metabolism have also been identified.[36,120-122,128,136] Genes encoding the major transition metal transporters of *C. reinhardtii* have also been characterized.[137,138]

Many of the genes of *C. reinhardtii* encode proteins similar to those present in animal cells. Recently, a comparison of *C. reinhardtii* gene models (generated by the Joint Genome Institute), with polypeptides encoded on the human genome, generated 4,348 matches (E value of 10^{-10} or below).[109] In *C. reinhardtii*, the basal bodies and flagella are similar to animal cell centrioles and cilia, respectively, which are not present in plant cells; centrioles are required for cilia assembly. Of the 4,348 'matched' polypeptides encoded on the *C. reinhardtii* and human genomes, 688 did not match any predicted polypeptides encoded on the *A. thaliana* genome. This subset of putative polypeptides included many of the approximately 150 and 250 polypeptides required for the biogenesis of basal bodies and flagella, respectively.[139,140] In addition to the known flagellar and basal body polypeptides, this group of genes encoded some proteins associated with human diseases and the impairment of cilia or basal body function. For example, there were six genes (BBS1, 2, 4, 5, 7 and 8) linked with Bardet-Biedl syndrome in humans, which can cause retinal dystrophy, obesity, polydactyly, renal and genital malformation and learning disabilities. Using RNAi technology, Dutcher and coworkers demonstrated a role for the BBS5 protein of *C. reinhardtii* in the assembly/function of the basal body, demonstrating how a relatively simple system may be used to elucidate the basis of some human diseases.

The generation of large genomic and cDNA databases has provided the foundation for developing methods to examine global gene expression and the proteome of *C. reinhardtii*. Both cDNA and oligonucleotide-based microarrays have been developed for *C. reinhardtii*, and they have been used to examine regulation of genes encoding polypeptides involved in nutrient assimilation, nutrient deprivation responses, light dependent responses[116,120-122,125] and the acclimation of cells to excess light and the generation of reactive oxygen radicals.[119] Microarrays and macroarrays have also been exploited to identify genes controlled by specific regulatory elements including CCM1 or CIA5, which controls responses of *C. reinhardtii* to inorganic carbon (Ci) starvation,[118,122] SAC1, which controls the response of the cell to sulfur starvation,[116] PSR1, which controls the responses of the cells to phosphorus starvation,[120] and PHOT, which controls the responses of the cell to blue light.[125] Nearly all genes induced under low Ci conditions that encode components of the carbon concentrating mechanism are controlled by CCM1/CIA5,[118,122] while SAC1 appears to control most sulfur assimilation genes as well as a subset of genes encoding proteins associated with oxidative stress and restructuring of the photosynthetic apparatus.[116] In addition, the *npq1 lor1* mutant, defective in the synthesis of photoprotective carotenoids, is more sensitive to reactive oxygen species.[119]

Genomic information is an important resource for those using proteomic approaches. This information has allowed Hippler and colleagues[126] to identify specific light harvesting polypeptides on two dimensional gels and also demonstrate specific amino-terminal processing of the LHCBM3 and LHCBM6, two polypeptides associated with photosystem II light harvesting. The identification of proteins associated with a complex that binds and regulates specific mRNAs in a circadian manner was also facilitated by genomic sequence information.[123,141] Recently, *C. reinhardtii* proteome projects have also been developed (http://www.uni-bielefeld.de/biologie/Zellphysiologie/kruse/Systematicproteomics.html; http://www.uni-jena.de/content_lang_en_page_5929.html; http://www-dsv.cea.fr/content/cea_eng>/d_dep/d_drdc/ d_ cp/).

Thalassiosira pseudonana

Diatoms represent a large and diverse group of organisms populating marine, freshwater and terrestrial environments. These organisms may be responsible for as much as 20% of global primary productivity. Generally, diatoms have a centric or pennate structure, but may also exhibit more novel morphologies (e.g., *Phaeodactylum tricornutum* has a morphotype resembling a three cornered hat). Perhaps the most characteristic visual feature of diatoms is their intricately patterned and often ornate, silicified cell walls.

Although still early in development, molecular tools have been tailored for use with diatoms. These organisms can be transformed and the DNA introduced into the cells is randomly integrated into the genome.[142-144] Reporter genes are now available, which include the *E. coli uidA* gene (encodes β-glucuronidase), the *cat* gene (encodes chloramphenicol acetyl transferase), the firefly *luc* gene (encodes luciferase),[145] a variant of the gene encoding green fluorescent protein (*egfp*) and the jellyfish aequorin gene[146] (see also Chapter 3). The targeting of proteins to subcellular compartments in the diatom *P. tricornutum* has been examined using GFP fusion proteins in conjunction with confocal microscopy.[147,148] Interestingly, a gene encoding the human glucose transporter fused to GFP and introduced into *P. tricornutum* was shown to be integrated into the cytoplasmic membranes. The chimeric transporter was able to function in the diatom, and the transformed *P. tricornutum*, normally an obligate photoautotrophic alga, was able to grow as a heterotroph (in the dark using exogenous glucose).[148] At this point, one of the most serious drawbacks in working with diatoms stems from the fact that they are diploid and that no procedures have been developed to consistently achieve sexual crosses. The inability to mate these organisms and generate progeny makes it nearly impossible to generate populations of recessive mutants and to isolate the mutated gene. It may take a strong concerted effort to understand the factors that control the diatom life cycle and to be able to control those factors.[149-155]

Diatoms were developed for genomic studies based on a number of their characteristics: (1) they are ecological relevant organisms, important as primary Ci fixers and in the biogeochemical cycling of nutrients, (2) they make a highly patterned cell wall that might contribute to developments in nanotechnology, (3) they are readily manipulated at the molecular level and (4) some have relatively small genomes. Little molecular analyses have been performed with diatoms and most information with respect to genome size comes from studies of Veldhuis et al;[156] in this study the sizes of seven diatom genomes were estimated based on DNA staining using PicoGreen or SYTOX Green; the genomes varied in size from 34 to ~700 Mb. The diatoms initially chosen for genome sequencing were the centric diatom *Thalassiosira pseudonana* and the pennate diatom *Phaeodactylum tricornutum. T. pseudonana* is a small, silicified diatom with a genome of a little over 30 Mb. Some other representatives of the genus *Thalassiosira* are considered to be ecologically more important, especially *T. weissflogii*, however, the latter has a genome that is estimated to be nearly 20 times larger than that of *T. pseudonana*. This finding raises the interesting question of why one organism, which has been classified as in the same genera, as the other has so much more DNA than the other. The *T. pseudonana* strain ultimately used for genome sequencing was collected from Moriches Bay (Long Island) in 1958 {CCMP1335; Center for Culture of Marine Phytoplankton; (http://ccmp.bigelow.org/)}. The *T. pseudonana* nuclear genome sequence was completed in 2004[30] by the Joint Genome Institutes or JGI (http://genome.jgi-psf.org/thaps1/thaps1.home.html). Its genome is about ~34 Mb, which is distributed over 24 chromosomes ranging in size from 0.66-3.32 Mb. The genome is predicted to encode ~11,242 polypeptides and the organism has metabolic pathways important for both photoautotrophic and photoheterotropphic growth.

Diatoms have interesting life cycles, metabolic pathways and morphologies, and the analysis of the genome of *T. pseudonana* has and will continue to provide insights into various aspects of the biology of these organisms. A major area of interest in diatom biology and nanotechnology concerns the biogenesis and organization of the diatom frustule or cell wall, which is silicified and highly patterned. There has been considerable interest concerning the ways in which the cells acquire, package and assemble silica.[157-159] While understanding diatom cell wall biogenesis and developing ways to alter silica deposition in vivo will contribute to novel silicon-based nano-fabrication technologies, there are numerous fundamental biological questions associated with diatom cell wall biogenesis. For example, how and where are the different wall components assembled, what cellular factors dictate pattern configuration, how is the pattern decorated with spins, knobs etc, once the wall is fully assembled, how can we use genetic and molecular tools to modify wall architecture, and what will it tell us about the biogenesis of the structure and the ways in which intricate patterns are established? Initial studies focusing on deposition of cell wall material noted that silica polymerization appeared to occur in intracellular vesicles called silica deposition vesicle (SDV).[160,161] The sites of silicifaction and features of deposition and assembly appear to be controlled primarily by cytoskeletal components, including microtubules and actin.[162,163] The siliffins, polyanionic phosphoproteins integral to diatom cell walls,[164-168] appear to play an important role in deposition and patterning of cell wall silica. Analysis of the genome of *T. pseudonana* has revealed the presence of 5 genes encoding silaffins; the different silaffin polypeptides are likely to have distinct structural/functional associations within the wall. Furthermore, long chain polyamines associated with the cell wall may function in the polymerization of the silica.[165,169] Other wall-associated proteins include the frustulins, glycoproteins that are probably not directly involved in assembly of silica building blocks for frustule construction.[170]

Relevant to both metabolism and cell wall biosynthesis in diatoms was the finding that these organisms have the genetic potential to synthesize enzymes of the urea cycle. This cycle, most likely localized to the mitrochondrion, allows these organisms to use urea as a sole source of nitrogen, which may reflect the fact that significant levels of urea are present in marine environments. The urea cycle may also facilitate the synthesis of the high energy intermediate creatine phosphate, which can drive a number of cellular processes. One intermediate in the urea cycle is ornithine, a precursor in spermine and spermidine synthesis.[171,172] Interestingly, there are a number of copies of genes on the *T. pseudonana* genome involved in spermine and spermidine synthesis and these small polyamines are likely the precursors for the longer polyamine molecules that are present in the cell wall.

The genome sequence of *T. pseudonana* has brought into focus a number of interesting aspects of diatom biology. Some of these issues involve the ways in which diatoms position themselves in the water column, the function, structure and evolution of light harvesting components,[173,174] mechanisms associated with the dissipation of excess absorbed light energy,[175-177] the role of the C4 pathway in CO_2 fixation,[178] the biosynthesis of long chain polyunsaturated fatty acids,[179-182] the role of Ca^{++} in signaling processes,[146] the identification and specific functions of specific photoreceptors, and the ways in which sexuality and cell morphogenesis are controlled.

The position of some diatoms in the water column is regulated by availability of light and nutrients. There are several mechanisms by which algae may change their buoyancy and position within the water column. One interesting mechanism involves the extrusion of chitin fibers through the pores of the frustule.[183] The genome of *T. pseudonana* has numerous genes that encode polypeptides with potential roles in the biosynthesis and degradation of chitin. These genes may be controlled by environmental factors, and their activity may in turn regulate chitin extrusion and the depth-position of the diatom in the water column.

Diatoms have a xanthophyll cycle in which diadinoxanthin is converted to diatoxanthin; this cycle contributes to the capacity of the organism to dissipate excess absorbed light energy.[175,177] In vascular plants, the PsbS protein is critical for xanthophyll cycle-dependent energy dissipation, which is measured as nonphotochemical quenching of fluorescence.[133,134,184-186] Interestingly, neither the *P. tricornutum* nor *T. pseudonana* genomes have a gene encoding PstS. The quenching of absorbed light energy in the diatoms can be very pronounced (dropping well below Fo) and it will be important to elucidate the polypeptides of these organisms that are critical for nonphotochemical quenching. *T. pseudanana* also has no identified genes encoding light harvesting-related, stress-associated ELIPs and SEPs, although there are two genes encoding the related HLIPs.

The C4 pathway appears to play a role in the fixation of inorganic carbon in the diatom *T. weissflogii.*[178] While many enzymes involved in C4 metabolism are encoded on the *T. pseudonana* genome, an enzyme capable of decarboxylation of C4 acids in the plastid to generate CO_2, the substrate of ribulose-1,5-bisphosphate carboxylase, was not identified. Diatoms may have novel or highly diverged enzymes that functions in this capacity. Other areas of immediate interest concern the synthesis of commercially valuable, long chain polyunsaturated fatty acids (eicosapentaenoic and docosahexaenoic acids) and ways in which specific photoreceptors (at least phytochrome and cryptochrome) and nutrient acquisition systems are linked to diatom ecology.

Recently, JGI has sequenced the genome of *P. tricornutum*. Completion of and public access to this sequence will allow a comparison between centric and pennate species and may also help us understand the genetic basis of morphotype differentiation. *P. tricornutum* is not considered to be very ecologically important (it is considered an 'atypical' diatom), but a number of molecular tools including reporter genes, selectable markers and a transformation system[143,14] have been developed for this organism. Furthermore, its genome is very small (approximately 20 Mb), and there is an abundant literature on its morphology, physiology and ecology. There is also a relatively large-scale EST project, and a queryable database (PtDB version1.0, http://www.szn.it/plant/PhaeodactylumEST/home.htm) that is helping in the analysis of the genomic sequences.

Cyanidioschyzon merolae

Cyanidioschyzon merolae, a red algae that is a member of the Cyanidiales, grows in hot springs (45°C, pH of 1.5) and is considered a primitive eukaryotic alga.[29,187,188] This group includes the genera *Cyanidium*, *Cyanidioschyzon* and *Galdieria*, with recent findings that demonstrate an unexpectedly high level of genetic diversity among the Cyanidiales.[189] *C. merolae* has a single golgi apparatus, a simple system of internal membranes, and can be transformed; the exogenous DNA, introduced by electroporation, may integrate into the nuclear genome by homologous recombination.[190] Furthermore, the chloroplast genome is ~150 kbp and contains 243 genes,[191] many of which are overlapping (~40%). This overlapping gene arrangement yields a highly compacted plastid genome. A major focus of the research with *C. merolae* has explored mechanisms by which mitochondria and plastids divide.[187,192-200]

C. merolae was the first alga for which a full nuclear genome sequence was reported.[201] This genome, approximately 16.5 Mb with 5,331 genes, is distributed over 20 chromosomes and contains three distinct rDNA loci.[201] The primitive nature of *C. merolae* is reflected by the finding that the nucleolus is a 'minimal' nucleolus since it is small and devoid of chromatin, the small genome contains only 26 genes that have introns, and the gene families that are present have low member representation; good examples of this are the gene families encoding motor proteins. There are genes for one set of tubulin subunits, two actins and both intermediate filament and kinesin family proteins. There are no genes encoding myosin. The dynamins, which function in the division of mitochondria and chloroplasts, are represented by two genes, whereas in most organisms they are part of a family with far more gene members. The reduced set of cytokinesis- and cell motility-associated genes, in accord with the minimum gene content, might reflect the specialized environment in which this organism functions.

C. merolae genomic sequences may reveal aspects of the endosymbiont origins of plastids and plastid function. Enzymes of the reductive pentose monophosphate cycle, involved in the fixation of Ci, appear to have a mosaic composition; some genes encoding enzymes in the cycle appear to be derived from the presumed cyanobacterial endosymbiont and others from the eukaryotic host. The origin of the RPM genesin of *C. merolae* and *A. thaliana* is similar, suggesting that they were derived from a common ancestral organism and that the genetic composition of the system has remained stable after the two lineages separated. Other interesting observations include, the tRNAs contain ectopic introns, there are no genes encoding phototropins (blue UV-A light photoreceptors) and phytochromes (red light photoreceptors), and the nuclear genome has a single histidine kinase and no response regulators. Again, the limited number of signaling elements encoded on the *C. merolae* nuclear genome may be indicative of the highly specialized adaptation to environments in which this organism grows.

Nucleomorph Genome of **Guillardia theta**

Like other chlorophyll c-containing chromophytic algae, the Cryptomonads have evolved as a consequence of the engulfment of a red alga by a eukaryotic protist. However, unlike most of the chromophytes, the Cryptomonads are unique in that they have an enslaved, diminutive red algal nucleus that has remained after the secondary endosymbiotic event.[202,203] The reduced red algal nucleus or nucleomorph is delineated by an envelop membrane with nuclear pores. There are three highly reduced chromosomes in the nucleomorph of *Guillardia theta*, which together constitute 551 kbp. The nucleomorph genome is tightly compacted, with almost no noncoding DNA and a predicted coding capacity of 464 polypeptides, about half of which have no known function. Of the protein encoding genes, only 17 have introns, and 11 of these are for ribosomal proteins.

A number of findings with respect to the protein encoding capacity of the nucleomorph raise important questions. Most nucleomorph-encoded proteins are required for replication of chromosomes, gene expression, biogenesis of periplastid ribosomes, with a few being involved in other cellular functions. While some nucleomorph-encoded polypeptides participate in mRNA processing, removal of tRNA introns and maturation of rRNA, there are also genes encoding 30 plastid-targeted proteins, 3 transporters, and a few enzymes with anabolic and regulatory functions. Since the plastid genome only has a small fraction of the genes required for plastid function, and the nucleomorph only encodes an additional 30 chloroplast localized proteins, most of polypeptides that function in the plastids must be synthesized in the cytoplasm of the cell, traverse the rough ER, the periplastid membrane, pass through the periplastid space, and cross the double envelop plastid membrane to reach their site of function within the organelle. The arrangement of the nucleomorph with respect to other cellular membranes of the cell is presented by Douglas et al.[21]

Of the nucleomorph-encoded, plastid-targeted polypeptides, only a few function in photosynthesis {rubredoxin and HLIP; the latter are small polypeptides of the light harvesting gene family that may be critical for the management of excitation energy},[204] organelle division, gene expression, nucleic acid metabolism and protein translocation into plastids and thylakoid membranes. Nucleomorph-encoded plastid polypeptides have amino terminal presequences that, in the case of rubredoxin, can function as a transit peptide.[205,206] Other nucleomorph genes encode regulatory proteins and RNA polymerase subunits, factors involved in translation and starch accumulation, histones, histone acetylase, and proteins involved in proteolysis and DNA replication. Some polypeptides essential for nucleomorph functions, including subunits of DNA polymerase, are not encoded by the nucleomorph genome and are likely to be synthesized in the cytoplasm of the cell and then transported into the nucleomorph.

Delineating the events critical for the biogenesis of the plastid and for the maintenance of the nucleomorph and periplastid compartment would help elucidate ways in which gene expression is coordinated within different subcellular compartments, the signals that pass between the compartments, and the mechanisms involved in routing metabolites and targeting proteins to specific organelles. Furthermore, sequencing the plastid, nucleomorph and nuclear genomes in the the Cryptomonads will define the genetic content of each of the cellular gene repositories, and perhaps provide insights into mechanisms involved in the transfer of nucleic acid among genomes and the steps involved in the conversion of an endosymbiont to an organelle.

cDNA and Partial Genome Sequences

Full-genome sequences are being generated for a number of diverse algae, and the list is growing every day. The organisms currently being sequenced are the multicellular green alga *Volvox carteri*, which is a close relative of *C. reinhardtii*, the diatom *P. tricornutum*, the planktonic bloom former *Aureococcus anophagefferens*, and the prasinophytes *Micromonas pusilla* and the picoeukaryote *Ostreococcus tauri*. *O. tauri* a primitive green alga with 18 chromosomes and a total genome size of ~11.5 Mb. A 7-fold sequence coverage of the genome has been completed, and the analysis of this sequence has led to the annotation of 4,000 ORFs; genomic information for *Ostreococcus tauri* has recently become available (http://bioinformatics.psb.ugent.be/genomes/browse/gbrowse/ostreococcus).

The approximately 220 Mb genome of *Emiliania huxleyi*, a coccolithophore, is also being sequenced. Coccolithophores generate calcium carbonate crystals that form defined structures of various shapes from the rhombohedral, to highly decorated plates and elaborate structures that can be shaped like coronets.[207,208] The development of the coccolith initiates in specific vesicles derived from the golgi; these coccoliths are exported from the cell and remain associated with the cell surface[209] as shown in Figure 1. There is an intrinsic interest in these organisms as they are of ecological importance and can form massive blooms that extend over 1000s of square kilometers of the ocean surface.[210,211] They are also important in the biomineralization and cycling of nutrients in the environment, and may help unveil mysteries associated with the ways in which cells pattern both internal and external structures. Recently, EST libraries have been constructed and partially characterized from *E. huxleyi*,[212,213] and libraries of differentially-expressed genes were constructed in order to elucidate enzymes and processes associated with biomineralization.[214] To date, only the GPA protein (a protein rich in glycine, proline and alanine that has a Ca^{2+}-binding motif) has been associated with the formation of coccoliths. Coccolith formation has also been linked to nutrient deprivation conditions, especially phosphate limitation.[215,216] To explore phosphate limitation responses and the formation of coccoliths, *E. huxleyi* grown in nutrient-replete medium was exposed to medium lacking P, conditions that would trigger calcification. Of the sequences specifically expressed during P starvation, one showed similarity to ligand-gated anion channels, although the role of this channel in phosphate utilization or biomineralization is not clear. Interestingly, a large number of the differentially-expressed sequences have no homology with other sequences in the databases; some of these novel proteins may be associated with biomineralization.[214]

A number of algae are also being used for the generation and sequencing of cDNAs. A cDNA library has been constructed from RNA isolated from the dinoflagellate *Alexandrium tamarense* and 3,628 unique cDNAs identified (see http://genome.uiowa.edu/projects/dinoflagellate/). cDNA libraries have also been made for the dinoflagellates *Lingulodinium polyedrum* and *Amphidinium carterae*.[217] Interestingly, many genes normally present on the plastid genome in photosynthetic organisms have moved into the dinoflagellate nuclear genome.[217-219] Analyses of cDNA libraries of *Porphyra yezeonsis*[220] (also see http://www.kazusa.or.jp/en/plant/

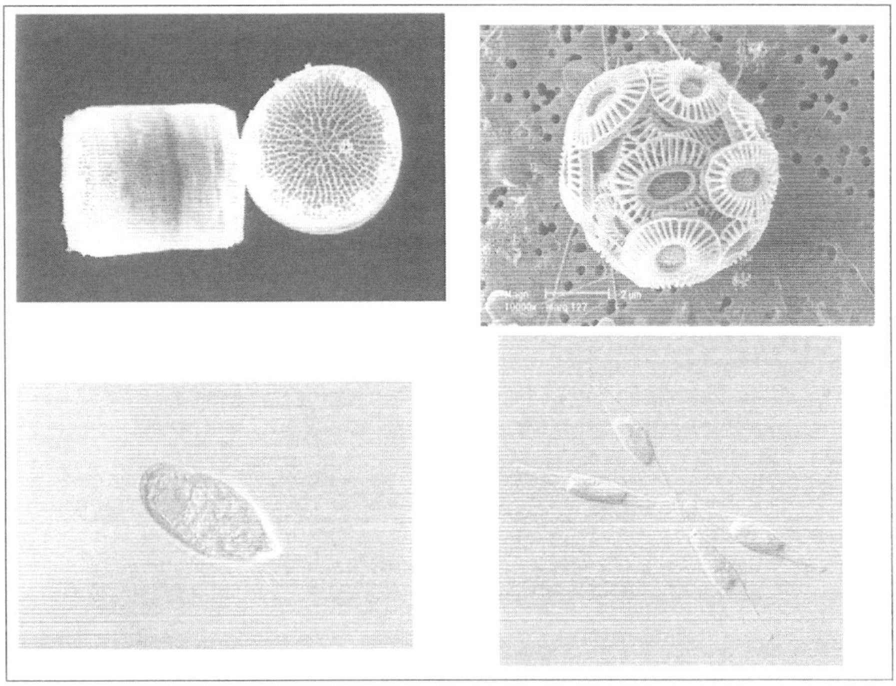

Figure 1. Images of two cells of the centric diatom *Thalassiosira pseudonana* (upper, left) from Nils Kröger (http://www.awi-bremerhaven.de/AWI/Presse/PM/pm04-2.hj/040929Gensequenz-e.html),the coccolithiphore Emiliania huxleyi (upper, right)(http://www.noc.soton.ac.uk/soes/staff/tt/eh/), originally generated by MarkusGeisen and Jeremy Young, NHM London, the small red alga *Cyanidioschyzon merolae* (lower, left)(http://www.nies.go.jp/biology/mcc/images/PCD4211/0198L.jpg)(www.nies.go.jp), and four cells of the pennate diatom *Phaeodactylum tricornutum* (lower, right)(http://biology.plosjournals.org/perlserv/?request=get-document&doi=10.1371/journal.pbio. 0020306)

porphyra/EST/), *Emiliania huxleyi*[212,213] *Phaeodactylum tricornutum*[221] and *Laminaria digitata*[222] have also begun. However, many of these projects are still in their early stages of development and a more thorough analysis of the sequence information that is being generated will add new insights into the biology of these organisms.

Viral Genomes

There are many algae and bacteria that can be infected by viruses, and there has been a strong interest in these viruses, especially some that have recently been identified in marine habitats. In the marine environment, viruses appear to be particularly abundant as estimated by transmission electron microscopy;[223-225] they reach titres as high as 10^7 virus particles per ml[226,227] and can function as pathogens of planktonic organisms. Marine phage represent a morphologically diverse group, based on TEM analyses of samples taken from the environment as well as isolated phage,[223,228] and capture extensive genetic diversity. They can cause substantial microbial mortality and are important for nutrient cycling in the oceans.[229] Many marine cyanophage harbor genes encoding proteins involved in photosynthesis, including *psbA* and *psbB*,[230-232] which might enable gene exchange among cyanobacteria and allow the phage to alter photosynthetic function in its host.[233-235] Furthermore, studies focused on defining the host range of marine phage reveal complex patterns of susceptibility.[236] Marine bacteria can be infected by the double-stranded, tailed myoviruses, podoviruses and siphoviruses, and while the field is still in its infancy, the eukaryotic plankton also appear to be infected by a range of viruses, including some viral categories that have not been previously characterized.

A rapidly increasing body of knowledge has demonstrated infection of marine eukaryotic algae by a variety of viral types. The bloom forming *Emiliania huxleyi* and the macroalgae *Ectocarpus siliculosus* are infected by double stranded Phycodnaviridae.[237-239] The haptophyte *Heterosigma akashiwo* was shown to be host for a single-stranded RNA virus.[240] A double-stranded RNA virus was shown to infect the prymnesiophyte *Micromonas pusilla.*[241] A nuclear-inclusion virus with both single- and double-stranded DNA forms infects the diatom *Chaetoceros salsugineum.*[242]

While algal virology is a nascent field and there is still much to be discovered with respect to phage diversity and mechanisms of infection (and potential ways to use these phage to deliver DNA to the algae), an increasing knowledge-base of sequences from double-stranded DNA viruses is beginning to reveal evolutionary relationships among viral groups.[243] There are families of large, double-stranded viruses, including Poxviridae, Iridoviridae, Asfarviridae, Phycodnaviridae and Mimiviridae, that are likely to have a common ancestor.[244] The Phycodnaviridae represents a diverse, ancient family of double-stranded DNA viruses with an icosahedral morphology and an internal lipid membrane. These viruses infect numerous eukaryotic algae (primary producers of the oceans) and could have genomes >500 kbp.[245,246] The two viruses in this family that were first sequenced were the *Paramedium bursaria chlorella virus* (PBCV-1)[247] and the *Ectocarpus siliculosus virus* (ESV-1),[239] although there are many other phaeoviruses that infect filamentous brown algae.[248] There appears to be extensive genetic variation within members of this viral family, present in marine environments, as determined from sequences of DNA polymerase genes.[249] Similar viral sequences are found in oceans separated by thousands of miles,[250] and recent studies have demonstrated the occurrence of extensive genetic diversity in a number of the viral families present in coastal oceans and sediments; there appear to be many thousand viral genotypes.[251]

Populations of *E. siliculosus* present in all temperate coastal regions are infected with the EsV-1 type phaeoviruses; the size of this phaeovirus particle ranges from 1300 to 1800 Angstroms. The double-stranded genomes of both PBCV-1 and EsV-1 are larger than 330 kbp. EsV-1 appears to have a linear genome of which the termini are inverted repeats with short nonidentical regions.[239] PBCV-1 also has a linear genome with ends that form terminal inverted repeats, but these repeats are 100% identical and each end forms a covalently closed hairpin. The EsV-1 genome is predicted to encode 231 polypeptides while the PBCV-1 genome may encode 375 polypeptides. The encoded proteins are a mixture of eukaryote- and prokaryote-like polypeptides that represent activities involved in many cellular processes including DNA replication, transcription and translation, ion transport, nucleotide, sugar and lipid metabolism, proteins glycosylation and cell wall degradation. Only 33 of the encoded polypeptides are common to the two genomes.[238] Interestingly, the genomes of neither PBCV-1 nor EsV-1 encode RNA polymerase subunits. Five proteins encoded by both of the viral genomes function in DNA replication and recombination, although neither genome encodes a full complement of polypeptides required for DNA replication (e.g., both lack a gene encoding DNA primase). The *Chlorella* and brown algal viruses have a number of other important physiological distinctions. For example, while the *Chlorella* virus gains entry to its host by first digesting the cell wall at the viral attachment point, EsV-1 binds to the plasma membrane of host cells that lack a cell wall (e.g., some spores or gametes). Furthermore, while EsV-1 can be lysogenic, there is no stage in the life cycle of PBVC-1 where it becomes lysogenic. Interestingly, the coccolithovirus EhV-86, which infects *Emiliania huxleyi*, encodes its own RNA polymerase (unlike most other family members).[252-25] This coccolithovirus has 472 putative coding sequences and surprisingly, only 14% resemble other sequences present in the databases.

Concluding Remarks

Morphological, physiological and molecular aspects of many of the algal groups are fascinating, and often markedly different from those of land plants. Currently, there are a relatively small number of algae for which full or nearly full genome sequences are available, but this is likely to change as the focus on environmental issues increases and as changing biosphere conditions impact on terrestrial and aquatic ecosystems. Furthermore, it will also be critical to understand how the dense and diverse viral populations in the oceans shape both functional and evolutionary aspects of algal genomes.

Acknowledgements

I thank NSF for supporting genomic research using *Chlamydomonas reinhardtii* and awarding us grant MCB 0235878. I also thank Dan Rokhsar and Igor Grigoriev at the Joint Genome Institutes, and members of the *C. reinhardtii* Consortium involved in developing the tools and infrastructure for securing and examining *C. reinhardtii* cDNA and genomic information, and for providing stimulating discussions and valuable insights. This is a Carnegie Institution Publication No. 1683.

References

1. Biegala IC, Not F, Vaulot D et al. Quantitative assessment of picoeukaryotes in the natural environment by using taxon-specific oligonucleotide probes in association with tyramide signal amplification-fluorescence in situ hybridization and flow cytometry. Appl Environ Microbiol 2003; 69(9):5519-29.
2. Díez B, Pedros-Alio C, Massana R. Study of genetic diversity of eukaryotic picoplankton in different oceanic regions by small-subunit rRNA gene cloning and sequencing. Appl Environ Microbiol 2001; 67(7):2932-41.
3. Graham LE, Wilcox LW. Algae. Upper Saddle River: Prentice Hall, 2000.
4. Berteau O, Mulloy B. Sulfated fucans, fresh perspectives: Structures, functions, and biological properties of sulfated fucans and an overview of enzymes active toward this class of polysaccharide. Glycobiology 2003; 13(6):29R-40R.
5. Feizi T, Mulloy B. Carbohydrates and glycoconjugates. Glycomics: The new era of carbohydrate biology. Curr Opin Struct Biol 2003; 13(5):602-4.
6. Drury JL, Dennis RG, Mooney DJ. The tensile properties of alginate hydrogels. Biomaterials 2004; 25(16):3187-99.
7. Matsubara K. Recent advances in marine algal anticoagulants. Curr Med Chem Cardiovasc Hematol Agents 2004; 2(1):13-9.
8. Chamberlain JG. The possible role of long-chain, omega-3 fatty acids in human brain phylogeny. Perspect Biol Med 1996; 39(3):436-45.
9. Salem Jr N, Moriguchi T, Greiner RS et al. Alterations in brain function after loss of docosahexaenoate due to dietary restriction of n-3 fatty acids. J Mol Neurosci 2001; 16(2-3):299-307.
10. Murdoch L. Discovering the Great Barrier Reef. Sydney: Harper Collins, 1996.
11. Coles SL, Brown BE. Coral bleaching—Capacity for acclimatization and adaptation. Adv Mar Biol 2003; 46:183-223.
12. Beja O, Spudich EN, Spudich JL et al. Proteorhodopsin phototrophy in the ocean. Nature 2001; 411(6839):786-9.
13. de la Torre JR, Christianson LM, Beja O et al. Proteorhodopsin genes are distributed among divergent marine bacterial taxa. Proc Natl Acad Sci USA 2003; 100(22):12830-5.
14. Venter JC, Remington K, Heidelberg JF et al. Environmental genome shotgun sequencing of the Sargasso Sea. Science 2004; 304(5667):66-74.
15. Whitelaw CA, Barbazuk WB, Pertea G et al. Enrichment of gene-coding sequences in maize by genome filtration. Science 2003; 302(5653):2118-20.
16. Mayer K, Mewes HW. How can we deliver the large plant genomes? Strategies and perspectives. Curr Opin Plant Biol 2002; 5(2):173-7.
17. Turmel M, Otis C, Lemieux C. The complete chloroplast DNA sequence of the green alga Nephroselmis olivacea: Insights into the architecture of ancestral chloroplast genomes. Proc Natl Acad Sci USA 1999; 96(18):10248-53.

18. Turmel M, Otis C, Lemieux C. The chloroplast and mitochondrial genome sequences of the charophyte Chaetosphaeridium globosum: Insights into the timing of the events that restructured organelle DNAs within the green algal lineage that led to land plants. Proc Natl Acad Sci USA 2002; 99(17):11275-80.
19. Wakasugi T, Nagai T, Kapoor M et al. Complete nucleotide sequence of the chloroplast genome from the green alga Chlorella vulgaris: The existence of genes possibly involved in chloroplast division. Proc Natl Acad Sci USA 1997; 94(11):5967-72.
20. Lemieux C, Otis C, Turmel M. Ancestral chloroplast genome in Mesostigma viride reveals an early branch of green plant evolution. Nature 2000; 403(6770):649-52.
21. Douglas S, Zauner S, Fraunholz M et al. The highly reduced genome of an enslaved algal nucleus. Nature 2001; 410(6832):1091-6.
22. Tada N, Shibata S, Otsuka S et al. Comparison of gene arrangements of chloroplasts between two centric diatoms, Skeletonema costatum and Odontella sinensis. DNA Seq 1999; 10(4-5):343-7.
23. Chu KH, Qi J, Yu ZG et al. Origin and phylogeny of chloroplasts revealed by a simple correlation analysis of complete genomes. Mol Biol Evol 2004; 21(1):200-6.
24. Stirewalt VL, Michalowski CB, Loffelhardt W et al. Nucleotide sequence of the cyanelle genome from Cyanophora paradoxa. Plant Mol Biol 1995; 13:327-332.
25. Glockner G, Rosenthal A, Valentin K. The structure and gene repertoire of an ancient red algal plastid genome. J Mol Evol 2000; 51(4):382-90.
26. Hallick RB, Hong L, Drager RG et al. Complete sequence of Euglena gracilis chloroplast DNA. Nucleic Acids Res 1993; 21(15):3537-44.
27. Zhang Z, Green BR, Cavalier-Smith T. Single gene circles in dinoflagellate chloroplast genomes. Nature 1999; 400(6740):155-9.
28. Zhang Z, Cavalier-Smith T, Green BR. Evolution of dinoflagellate unigenic minicircles and the partially concerted divergence of their putative replicon origins. Mol Biol Evol 2002; 19(4):489-500.
29. Matsuzaki M, Misumi O, Shin IT et al. Genome sequence of the ultrasmall unicellular red alga Cyanidioschyzon merolae 10D. Nature 2004; 428(6983):653-7.
30. Armbrust EV, Berges JA, Bowler C et al. The genome of the diatom Thalassiosira pseudonana: Ecology, evolution, and metabolism. Science 2004; 306(5693):79-86.
31. Debuchy R, Purton S, Rochaix JD. The argininosuccinate lyase gene of Chlamydomonas reinhardtii: An important tool for nuclear transformation and for correlating the genetic and molecular maps of the ARG7 locus. EMBO J 1989; 8:2803-2809.
32. Kindle KL, Schnell RA, Fernández E et al. Stable nuclear transformation of Chlamydomonas using the Chlamydomonas gene for nitrate reductase. J Cell Biol 1989; 109:2589-2601.
33. Diener DR, Curry AM, Johnson KA et al. Rescue of a paralyzed flagella mutant of Chlamydomonas by transformation. Proc Natl Acad Sci USA 1990; 87:5739-5743.
34. Mayfield SP, Kindle KL. Stable nuclear transformation of Chlamydomonas reinhardtii by using a C. reinhardtii gene as the selectable marker. Proc Natl Acad Sci USA 1990; 87:2087-2091.
35. Shimogawara K, Fujiwara S, Grossman A et al. High-efficiency transformation of Chlamydomonas reinhardtii by electroporation. Genetics 1998; 148(4):1821-8.
36. Fernandez E, Cardenas J. Genetic and regulatory aspects of nitrate assimilation in algae. Oxford University Press, 1989:101-24.
37. Kindle KL. High-frequency nuclear transformation of Chlamydomonas reinhardtii. Proc Natl Acad Sci USA 1990; 87:1228-1232.
38. Goldschmidt-Clermont M. Transgenic expression of aminoglycoside adenine transferase in the chloroplast: A selectable marker for site-directed transformation of Chlamydomonas. Nucleic Acids Res 1991; 19:4083-4089.
39. Nelson JAE, Savereide PB, Lefebvre PA. The CRY1 gene in Chlamydomonas reinhardtii: Structure and use as a dominant selectable marker for nuclear transformation. Mol Cell Biol 1994; 14:4011-4019.
40. Stevens DR, Rochaix JD, Purton S. The bacterial phleomycin resistance gene ble as a dominant selectable marker in Chlamydomonas. Mol Gen Genet 1996; 251:23-30.
41. Lumbreras V, Stevens DR, Purton S. Efficient foreign gene expression in Chlamydomonas reinhardtii mediated by an endogenous intron. Plant J 1998; 14(4):441-447.
42. Auchincloss AH, Loroch AI, Rochaix JD. The arginosuccinate lyase gene of Chlamydomonas reinhardtii: Cloning of the cDNA and its characterization as a selectable shuttle marker. Mol Gen Genet 1999; 261:21-30.
43. Kovar JL, Zhang J, Funke RP et al. Molecular analysis of the acetolactate synthase gene of Chlamydomonas reinhardtii and development of a genetically engineered gene as a dominant selectable marker for genetic transformation. Plant J 2002; 29(1):109-17.

44. Purton S, Rochaix JD. Complementation of a Chlamydomonas reinhardtii mutant using a genomic cosmid library. Plant Mol Biol 1994; 24:533-537.
45. Zhang H, Herman PL, Weeks DP. Gene isolation through genomic complementation using an indexed library of Chlamydomonas reinhardtii DNA. Plant Mol Biol 1994; 24:663-672.
46. Lefebvre PA, Silflow CD. Chlamydomonas: The cell and its genomes. Genetics 1999; 151(1):9-14.
47. Shrager J, Hauser C, Chang CW et al. Chlamydomonas reinhardtii genome project. A guide to the generation and use of the cDNA information. Plant Physiol 2003; 131(2):401-8.
48. Funke RP, Kovar JL, Weeks DP. Intracellular carbonic anhydrase is essential to photosynthesis in Chlamydomonas reinhardtii at atmospheric levels of CO_2. Demonstration via genomic complementation of the high-CO_2-requiring mutant ca-1. Plant Physiol 1997; 114(1):237-244.
49. Randolph-Anderson BL, Sato R, Johnson AM et al. Isolation and characterization of a mutant protoporphyrinogen oxidase gene from Chlamydomonas reinhardtii conferring resistance to porphyric herbicides. Plant Mol Biol 1998; 38:839-859.
50. Wykoff DD, Davies JP, Melis A et al. The regulation of photosynthetic electron transport during nutrient deprivation in Chlamydomonas reinhardtii. Plant Physiol 1998; 117(1):129-39.
51. Yildiz FH, Davies JP, Grossman A. Sulfur availability and the SAC1 gene control adenosine triphosphate sulfurylase gene expression in Chlamydomonas reinhardtii. Plant Physiol 1996; 112(2):669-75.
52. Palombella AL, Dutcher SK. Identification of the gene encoding the tryptophan synthase beta-subunit from Chlamydomonas reinhardtii. Plant Physiol 1998; 117(2):455-464.
53. Tam LW, Lefebvre PA. Cloning of flagellar genes in Chlamydomonas reinhardtii by DNA insertional mutagenesis. Genetics 1993; 135:375-384.
54. Davies JP, Yildiz F, Grossman AR. Mutants of Chlamydomonas with aberrant responses to sulfur deprivation. Plant Cell 1994; 6(1):53-63.
55. Davies J, Yildiz F, Grossman AR. Sac1, a putative regulator that is critical for survival of Chlamydomonas reinhardtii during sulfur deprivation. EMBO J 1996; 15:2150-2159.
56. Smith EF, Lefebvre PA. PF16 encodes a protein with armadillo repeats and localizes to a single microtubule of the central apparatus in Chlamydomonas flagella. J Cell Biol 1996; 132:359-370.
57. Koutoulis A, Pazour GJ, Wilkerson CG et al. The Chlamydomonas reinhardtii ODA3 gene encodes a protein of the outer dynein arm docking complex. J Cell Biol 1997; 137(5):1069-1080.
58. Smith EF, Lefebvre PA. PF20 gene product contains WD repeats and localizes to the intermicrotubule bridges in Chlamydomonas flagella. Mol Biol Cell 1997; 8:455-467.
59. Zhang D, Lefebvre PA. FAR1, a negative regulatory locus required for the repression if the nitrate reductase gene in Chlamydomonas reinhardtii. Genetics 1997; 146:121-133.
60. Asleson CM, Lefebvre PA. Genetic analysis of flagellar length control in Chlamydomonas reinhardtii: A new long-flagella locus and extragenic suppressor mutations. Genetics 1998; 148:693-702.
61. Davies JP, Yildiz FH, Grossman AR. Sac3, an Snf1-like serine/threonine kinase that positively and negatively regulates the responses of Chlamydomonas to sulfur limitation. Plant Cell 1999; 11(6):1179-90.
62. Wykoff DD, Grossman AR, Weeks DP et al. Psr1, a nuclear localized protein that regulates phosphorus metabolism in Chlamydomonas. Proc Natl Acad Sci USA 1999; 96(26):15336-41.
63. Vysotskaia VS, Curtis DE, Voinov AV et al. Development and characterization of genome-wide single nucleotide pholymorphism markers in the green alga Chlamydomonas reinhardtii. Plant Physiol 2001; 127:386-389.
64. Kathir P, LaVoie M, Brazelton WJ et al. Molecular map of the Chlamydomonas reinhardtii nuclear genome. Eukaryot Cell 2003; 2(2):362-79.
65. Schroda M, Vallon O, Wollman FA et al. A chloroplast-targeted heat shock protein 70 (HSP70) contributes to the photoprotection and repair of photosystem II during and after photoinhibition. Plant Cell 1999; 11(6):1165-1178.
66. Jeong BR, Wu-Scharf D, Zhang C et al. Suppressors of transcriptional transgenic silencing in Chlamydomonas are sensitive to DNA-damaging agents and reactivate transposable elements. Proc Natl Acad Sci USA 2002; 99:1076-1081.
67. Sineshchekov OA, Jung KH, Spudich JL. The rhodopsins mediate phototaxis to low- and high-intensity light in Chlamydomonas reinhardtii. Proc Natl Acad Sci USA 2002; 99:225-230.
68. Wilson NF, Lefebvre PA. Characterization of GSK3, a flagellar kinase with a putative role in the regulation of flagella length. Tenth International Chlamydomonas Conference 2002, (Abstract).
69. Davies JP, Weeks DP, Grossman AR. Expression of the arylsulfatase gene from the beta 2-tubulin promoter in Chlamydomonas reinhardtii. Nucleic Acids Res 1992; 20(12):2959-65.
70. Fuhrmann M, Oertel W, Hegemann P. A synthetic gene coding for the green fluorescent protein (GFP) is a versatile reporter in Chlamydomonas reinhardtii. Plant J 1999; 19(3):353-361.

71. Minko I, Holloway SP, Nikaido S et al. Renilla luciferase as a vital reporter for chloroplast gene expression in Chlamydomonas. Mol Gen Genet 1999; 262:421-425.
72. Mayfield SP, Franklin SE, Lerner RA. Expression and assembly of a fully active antibody in algae. Proc Natl Acad Sci USA 2003; 100(2):438-42.
73. Davies JP, Grossman AR. Sequences controlling transcription of the Chlamydomonas reinhardtii beta 2-tubulin gene after deflagellation and during the cell cycle. Mol Cell Biol 1994; 14(8):5165-74.
74. Quinn JM, Merchant S. Two Copper-responsive elements associated with the Chlamydomonas Cyc6 gene function as targets for transcriptional activators. Plant Cell 1995; 7:623-638.
75. Jacobshagen S, Kindle KL, Johnson CH. Transcription of CABII is regulated by the biological clock in Chlamydomonas reinhardtii. Plant Molecular Biology 1996; 31(6):1173-1184.
76. Ohresser M, Matagne RF, Loppes R. Expression of the arylsulphatase reporter gene under the control of the NIT1 promoter of Chlamydomonas reinhardtii. Curr Genet 1997; 31:264-271.
77. Villand P, Ericksson M, Samuelsson G. Regulation of genes by the environmental CO_2 level. Plant Physiol 1997; 114:258-259.
78. Fuhrmann M, Ferbitz L, Eichler-Stahlberg A et al. Promoter activity monitored by heterologous expression of Renilla reniformis luciferase in Chlamydomonas reinhardtii. Tenth International Chlamydomonas Conference 2002, (Abstract).
79. Komine Y, Kikis E, Schuster G et al. Evidence for in vivo modulation of chloroplast RNA stability by 3'-UTR homopolymeric tails in Chlamydomonas reinhardtii. Proc Natl Acad Sci USA 2002; 99:4085-4090.
80. Boynton JE, Gillham NW, Harris EH et al. Chloroplast transformation in Chlamydomonas with high velocity microprojectiles. Science 1988; 240:1534-1538.
81. Newman SM, Boynton JE, Gillham NW et al. Transformation of chloroplast ribosomal RNA in Chlamydomonas: Molecular and genetic characterization of integration events. Genetics 1990; 126:875-888.
82. Whitelegge JP, Koo D, Erickson J. Site-directed mutagenesis of the chloropolast psbA gene encoding the D1 polypeptide of photosystem II in Chlamydomonas reinhardtii changes at aspartate 170 affect the assembly of a functional water-splitting manganese cluster. In: Murata N, ed. Research in Photosynthesis. Dordrecht: Kluwer Academic Publishers, 1992:151-154.
83. Hong S, Spreitzer RJ. Nuclear mutation inhibits expression of the chloroplast gene that encodes the large subunit of ribulose-1,5-bisphosphate carboxylase/oxygenase. Plant Physiol 1994; 106(2):673-678.
84. Takahashi Y, Matsumoto H, Goldschmidt-Clermont M et al. Directed disruption of the Chlamydomonas chloroplast psbK gene destabilizes the photosystem II reaction center complex. Plant Mol Biol 1994; 24:779-788.
85. Hallahan BJ, Purton S, Ivison A et al. Analysis of the proposed Fe-Sx binding region in Chlamydomonas reinhardtii. Photosyn Res 1995; 46:257-264.
86. Webber AN, Su H, Binghma SE et al. Site-directed mutations affecting the spectroscopic characteristics and mid-point potential of the primary donor in photosystem I. Biochemistry 1996; 39:12857-12863.
87. Zhu G, Spreitzer RJ. Directed mutagenesis of chloroplast ribulose-1,5-bisphosphate carboxylase-oxygenase. Loop 6 substitutions complement for structual stability but decrease catalytic efficiency. J Biol Chem 1996; 271:18494-18498.
88. Fischer N, Setif P, Rochaix JD. Targeted mutations in the psaC gene of Chlamydomonas reinhardtii: Preferential reduction of FB at low temperature is not accompanied by altered electron flow from Photosystem I to ferredoxin. Biochemistry 1997; 36:93-102.
89. Lardans A, Gillham NW, Boynton JE. Site-directed mutations at residue 251 of the photosystem II D1 protein of Chlamydomonas that result in a nonphotosynthetic phenotype and impair D1 synthesis and accumulation. J Biol Chem 1997; 272:210-216.
90. Larson EM, O'Brien CM, Zhu G et al. Specificity for activase is changed by a Pro-89 to Arg substitution in the large subunit of ribulose-1,5-biosphosphate carboxylase-oxgenase. J Biol Chem 1997; 272:17033-17037.
91. Melkozernov AN, Su H, Lin S et al. Specific mutations near the primary donor in Photosystem I from Chlamydomonas reinhardtii alters the trapping time and spectroscopic properties of P700. Biochemistry 1997; 36:2898-2907.
92. Xiong J, Hutchinson RS, Sayre RT et al. Modification of the photosystem II acceptor side function in a D1 mutant (arginine-269-glycine) of Chlamydomonas reinhardtii. Biochimica et Biophysica Acta 1997; 1322:60-76.
93. Finazzi G, Furia A, Barbagallo RP et al. State transitions, cyclic and linear electron transport and photophosphorylation in Chlamydomonas reinhardtii. Biochimica et Biophysica Acta 1999; 1413(3):117-129.

94. Higgs DC, Shapiro RS, Kindle KL et al. Small cis acting sequences that specify secondary structures in a chloroplast mRNA are essential for RNA stability and translation. Mol Cell Biol 1999; 19:8479-8491.
95. Harris EH. The Chlamydomonas sourcebook. A Comprehensive Guide to Biology and Laboratory Use. San Diego: Academic Press, 1989.
96. Harris EH. Chlamydomonas as a model organism. Annu Rev Plant Physiol Plant Mol Biol 2001; 52:363-406.
97. Sager R. Genetic systems in Chlamydomonas. Science 1960; 132:1459-1465.
98. Gorman DS, Levine RP. Cytochrome f and plastocyanin: Their sequence in the photosynthetic electron transport chain of Chlamydomonas reinhardtii. Proc Natl Acad Sci USA 1966; 54:1665-1669.
99. Bennoun P, Levine RP. Detecting mutants that have impaired photosynthesis by their increased level of fluorescence. Plant Physiol 1967; 42:1284-1287.
100. Givan AL, Levine RP. The photosynthetic electron transport chain of Chlamydomonas reinhardtii. VII. Photosynthetic phosphorylation by a mutant strain of Chlamydomonas reinhardtii deficient in active P700. Plant Physiol 1967; 42:1264-1268.
101. Lavorel J, Levine RP. Fluorescence properties of wild-type Chlamydomonas reinhardtii and three mutant strains having impaired photosynthesis. Plant Physiol 1968; 43:1049-1055.
102. Levine RP. The analysis of photosynthesis using mutant strains of algae and higher plants. Annu Rev Plant Physiol 1969; 20:523-540.
103. Levine RP, Goodenough UW. The genetics of photosynthesis and of the chloroplast in Chlamydomonas reinhardii. Annu Rev Genet 1970; 4:397-408.
104. Moll B, Levine RP. Characterization of a photosynthetic mutant strain of Chlamydomonas reinhardi deficient in phosphoribulokinase activity. Plant Physiol 1970; 46:576-580.
105. Sato V, Levine RP, Neumann J. Photosynthetic phosphorylation in Chlamydomonas reinhardti. Effects of a mutation altering an ATP-synthesizing enzyme. Biochimica et Biophysica Acta 1971; 253:437-448.
106. Pazour GJ, Dickert BL, Vucica Y et al. Chlamydomonas IFT88 and its mouse homologue, polycystic kidney disease gene tg737, are required for assembly of cilia and flagella. J Cell Biol 2000; 151(3):709-18.
107. Snell WJ, Pan J, Wang Q. Cilia and flagella revealed: From flagellar assembly in Chlamydomonas to human obesity disorders. Cell 2004; 117(6):693-7.
108. Pennarun G, Bridoux AM, Escudier E et al. Isolation and expression of the human hPF20 gene orthologous to Chlamydomonas PF20: Evaluation as a candidate for axonemal defects of respiratory cilia and sperm flagella. Am J Respir Cell Mol Biol 2002; 26(3):362-70.
109. Li JB, Gerdes JM, Haycraft CJ et al. Comparative genomics identifies a flagellar and basal body proteome that includes the BBS5 human disease gene. Cell 2004; 117(4):541-52.
110. Grossman AR. Chlamydomonas reinhardtii and photosynthesis: Genetics to genomics. Curr Opin Plant Biol 2000; 3(2):132-7.
111. Grossman AR, Harris EE, Hauser C et al. Chlamydomonas reinhardtii at the crossroads of genomics. Eukaryot Cell 2003; 2(6):1137-50.
112. Dent RM, Han M, Niyogi KK. Functional genomics of plant photosynthesis in the fast lane using Chlamydomonas reinhardtii. Trends Plant Sci 2001; 6(8):364-71.
113. Dutcher SK. Chlamydomonas reinhardtii: Biological rationale for genomics. J Eukaryot Microbiol 2000; 47(4):340-9.
114. Lilly JW, Maul JE, Stern DB. The Chlamydomonas reinhardtii organellar genomes respond transcriptionally and post-transcriptionally to abiotic stimuli. Plant Cell 2002; 14(11):2681-706.
115. Im CS, Zhang Z, Shrager J et al. Analysis of light and CO(2) regulation in Chlamydomonas reinhardtii using genome-wide approaches. Photosynth Res 2003; 75(2):111-25.
116. Zhang Z, Shrager J, Jain M et al. Insights into the survival of Chlamydomonas reinhardtii during sulfur starvation based on microarray analysis of gene expression. Eukaryot Cell 2004; 3(5):1331-48.
117. Yoshioka S, Taniguchi F, Miura K et al. The novel Myb transcription factor LCR1 regulates the CO_2-responsive gene Cah1, encoding a periplasmic carbonic anhydrase in Chlamydomonas reinhardtii. Plant Cell 2004; 16(6):1466-77.
118. Miura K, Yamano T, Yoshioka S et al. Expression profiling-based identification of CO_2-responsive genes regulated by CCM1 controlling a carbon-concentrating mechanism in Chlamydomonas reinhardtii. Plant Physiol 2004; 135(3):1595-607.
119. Ledford HK, Baroli I, Shin JW et al. Comparative profiling of lipid-soluble antioxidants and transcripts reveals two phases of photo-oxidative stress in a xanthophyll-deficient mutant of Chlamydomonas reinhardtii. Mol Genet Genomics 2004; 272(4):470-9.

120. Moseley JL, Chang CW, Grossman AR. Genome-based approaches to understanding phosphorus deprivation responses and PSR1 control in Chlamydomonas reinhardtii. Eukaryot Cell 2006; 5(1):26-44.
121. Eberhard S, Jain M, Im CS et al. Generation of an oligonucleotide array for analysis of gene expression in Chlamydomonas reinhardtii. Curr Genet 2006; 49(2):106-24.
122. Wang Y, Sun S, Horken KM et al. Analyses of CIA5, the master regulator of the CCM in Chlamydomonas reinhardtii, and its control of gene expression. Can J Bot 2005; 83:765-779.
123. Wagner V, Fiedler M, Markert C et al. Functional proteomics of circadian expressed proteins from Chlamydomonas reinhardtii. FEBS Lett 2004; 559(1-3):129-35.
124. Wagner V, Gessner G, Mittag M. Functional proteomics: A promising approach to find novel components of the circadian system. Chronobiol Int 2005; 22(3):403-15.
125. Im CS, Eberhard S, Huang K et al. Phototropin involvement in expression of genes encoding chlorophyll and carotenoid biosynthesis enzymes and LHC apoproteins in Chlamydomonas reinhardtii. Plant J 2006, (In Press).
126. Stauber EJ, Fink A, Markert C et al. Proteomics of Chlamydomonas reinhardtii light-harvesting proteins. Eukaryot Cell 2003; 2(5):978-94.
127. Elrad D, Grossman AR. A genome's-eye view of the light-harvesting polypeptides of Chlamydomonas reinhardtii. Curr Genet 2004; 45(2):61-75.
128. La Fontaine S, Quinn JM, Nakamoto SS et al. Copper-dependent iron assimilation pathway in the model photosynthetic eukaryote Chlamydomonas reinhardtii. Eukaryot Cell 2002; 1(5):736-57.
129. Richly E, Leister D. NUPTs in sequenced eukaryotes and their genomic organization in relation to NUMTs. Mol Biol Evol 2004; 21(10):1972-80.
130. Savard F, Richard C, Guertin M. The Chlamydomonas reinhardtii LI818 gene represents a distant relative of the cabI/II genes that is regulated during the cell cycle and in response to illumination. Plant Mol Biol 1996; 32(3):461-73.
131. Richard C, Ouellet H, Guertin M. Characterization of the LI818 polypeptide from the green unicellular alga Chlamydomonas reinhardtii. Plant Mol Biol 2000; 42(2):303-16.
132. Gutman BL, Niyogi KK. Chlamydomonas and Arabidopsis. A dynamic duo. Plant Physiol 2004; 135(2):607-10.
133. Li XP, Bjorkman O, Shih C et al. A pigment-binding protein essential for regulation of photosynthetic light harvesting. Nature 2000; 403(6768):391-5.
134. Li XP, Gilmore AM, Caffarri S et al. Regulation of photosynthetic light harvesting involves intrathylakoid lumen pH sensing by the PsbS protein. J Biol Chem 2004; 279(22):22866-74.
135. Niyogi KK, Li XP, Rosenberg V et al. Is PsbS the site of nonphotochemical quenching in photosynthesis? J Exp Bot 2005; 56(411):375-82.
136. Bisova K, Krylov DM, Umen JG. Genome-wide annotation and expression profiling of cell cycle regulatory genes in Chlamydomonas reinhardtii. Plant Physiol 2005; 137(2):475-91.
137. Rosakis A, Koster W. Transition metal transport in the green microalga Chlamydomonas reinhardtii—genomic sequence analysis. Res Microbiol 2004; 155(3):201-10.
138. Rosakis A, Koster W. Divalent metal transport in the green microalga Chlamydomonas reinhardtii is mediated by a protein similar to prokaryotic Nramp homologues. Biometals 2005; 18(1):107-20.
139. Dutcher SK. Purification of basal bodies and basal body complexes from Chlamydomonas reinhardtii. Methods Cell Biol 1995; 47:323-34.
140. Dutcher SK. Flagellar assembly in two hundred and fifty easy-to-follow steps. Trends Genet 1995; 11(10):398-404.
141. Zhao B, Schneid C, Iliev D et al. The circadian RNA-binding protein CHLAMY 1 represents a novel type heteromer of RNA recognition motif and lysine homology domain-containing subunits. Eukaryot Cell 2004; 3(3):815-25.
142. Dunahey TG, Jarvis EE, Roessler PG. Genetic transformation of the diatoms Cyclotella cryptica and Navicula saprophila. J Phycol 1995; 31:1004-1012.
143. Apt KE, Kroth-Pancic PG, Grossman AR. Stable nuclear transformation of the diatom Phaeodactylum tricornutum. Mol Gen Gen 1996; 252:572-579.
144. Zaslavskaia LA, Lippmeier JC, Kroth PG et al. Transformation of the diatom Phaeodactylum tricornutum (Bacillariophyceae) with a variety of selectable marker and reporter genes. J Phycol 2000; 36:379-386.
145. Falciatore A, Casotti R, Leblanc C et al. Transformation of nonselectable reporter genes in marine diatoms. Mar Biotechnol 1999; 1(3):239-251.
146. Falciatore A, d'Alcala MR, Croot P et al. Perception of environmental signals by a marine diatom. Science 2000; 288(5475):2363-6.
147. Apt KE, Zaslavskaia LA, Lippmeier JC et al. In vivo characterization of diatom multipartite plastid targeting signals. J Cell Sci 2002; 115:4061-4069.

148. Zaslavskaia LA, Lippmeier JC, Shih C et al. Trophic conversion of an obligate photoautotrophic organism through metabolic engineering. Science 2001; 292:2073-2075.
149. Armbrust EV. Identification of a new gene family expressed during the onset of sexual reproduction in the centric diatom Thalassiosira weissflogii. Appl Environ Microbiol 1999; 65(7):3121-8.
150. Armbrust EV, Galindo HM. Rapid evolution of a sexual reproduction gene in centric diatoms of the genus Thalassiosira. Appl Environ Microbiol 2001; 67(8):3501-13.
151. Mann DG, Chepurnov VA, Droop SJM. Sexuality, incompatibility, size variation, and preferential polyendry in natural populations and clones of Sellaphora pupula (Bacilliarophyceae). J Phycol 1999; 35:152-170.
152. Mann DG. Patterns of sexual reproduction in diatoms. Hydrobiologia 1993; 269/270:11.
153. Armbrust EV, Chisholm SW. Role of light and cell cycle on the induction of spermatogenesis in a centric diatom. J Phycol 1990; 26:470-478.
154. Vaulot D, Olson RJ, Chisholm SW. Light and dark control of the cell cycle in two marine phytoplankton species. Exp Cell Res 1986; 167:38-52.
155. Vaulot D, Olson RJ, Merkel S et al. Cell cycle response to nutrient starvation in two phytoplankton species, Thalassiosira weissflogii and Hymenomonas carterae. Mar Biol 1987; 95:625-630.
156. Veldhuis MJW, Cucci TL, Sieracki ME. Cellular DNA content of marine phytoplankton using two new fluorophores: Taxonimic and ecological implications. J Phycol 1997; 33:527-541.
157. Hildebrand M, Volcani BE, Gassmann W et al. A gene family of silicon transporters. Nature 1997; 385(6618):688-9.
158. Hildebrand M, Dahlin K, Volcani BE. Characterization of a silicon transporter gene family in Cylindrotheca fusiformis: Sequences, expression analysis, and identification of homologs in other diatoms. Mol Gen Genet 1998; 260(5):480-6.
159. Hildebrand M, Wetherbee R. Components and control of silicification in diatoms. Prog Mol Subcell Biol 2003; 33:11-57.
160. Reimann BEF, Lewin JC, Volcani BE. Studies on the biochemistry and fine structure of silica shell formation in diatoms. II. The structure of the cell wall of Navicula pelliculosa (Breb.) Hilse. J Phycol 1966; 2:74-84.
161. Crawford RM, Schmid AMM. Ultrastructure of silica deposition in diatoms. In: Leadbeater BS, Riding R, eds. Biomineralization in Lower Plants and Animals. The Systematics Society, 1986.
162. Pickett-Heaps JD, Kowalski SE. Valve morphogenesis and the microtubule center of the diatom Hantzschia amphioxysis. Eur J Cell Biol 1981; 25:150-170.
163. Pickett-Heaps JD. Valve morphogenesis and the microtubule center in three species of the diatom Nitzschia. J Phycol 1983; 19:269-181.
164. Kröger N, Deutzmann R, Sumper M. Polycationic peptides from diatom biosilica that direct silica nanosphere formation. Science 1999; 286(5442):1129-32.
165. Kröger N, Deutzmann R, Bergsdorf C et al. Species-specific polyamines from diatoms control silica morphology. Proc Natl Acad Sci USA 2000; 97(26):14133-8.
166. Kröger N, Lorenz S, Brunner E et al. Self-assembly of highly phosphorylated silaffins and their function in biosilica morphogenesis. Science 2002; 298(5593):584-6.
167. Poulsen N, Sumper M, Kröger N. Biosilica formation in diatoms: Characterization of native silaffin-2 and its role in silica morphogenesis. Proc Natl Acad Sci USA 2003; 100(21):12075-80.
168. Poulsen N, Kröger N. Silica morphogenesis by alternative processing of silaffins in the diatom Thalassiosira pseudonana. J Biol Chem 2004, (In Press).
169. Lutz K, Groger C, Sumper M et al. Biomimetic silica formation: Analysis of the phosphate-induced self-assembly of polyamines. Phys Chem Chem Phys 2005; 7(14):2812-5.
170. Vrieling EG, Gieskes WWC, Beelen TPM. Silicon deposition in diatoms: Control by the pH inside the silicon deposition vesicle. J Phycol 1999; 35:548-559.
171. Morgan DM. Polyamines. An overview. Mol Biotechnol 1999; 11:229-250.
172. Igarashi K, Kashiwagi K. Polyamines: Mysterious modulators of cellular functions. Biochem Biophys Res Commun 2000; 271:559-564.
173. Oeltjen A, Marquardt J, Rhiel E. Differential circadian expression of genes fcp2 and fcp6 in Cyclotella cryptica. Int Microbiol 2004; 7:127-131.
174. Buchel C. Fucoxanthin-chlorophyll proteins in diatoms: 18 and 19 kDa subunits assemble into different oligomeric states. Biochemistry 2003; 42(44):13027-34.
175. Lavaud J, Rousseau B, van Gorkom HJ et al. Influence of the diadinoxanthin pool size on photoprotection in the marine planktonic diatom Phaeodactylum tricornutum. Plant Physiol 2002; 129(3):1398-406.
176. Lavaud J, Rousseau B, Etienne AL. Enrichment of the light-harvesting complex in diadinoxanthin and implications for the nonphotochemical fluorescence quenching in diatoms. Biochemistry 2003; 42(19):5802-8.

177. Lohr M, Wilhelm C. Algae displaying the diadinoxanthin cycle also possess the violaxanthin cycle. Proc Natl Acad Sci USA 1999; 96(15):8784-9.
178. Reinfelder JR, Milligan AJ, Morel FM. The role of the C4 pathway in carbon accumulation and fixation in a marine diatom. Plant Physiol 2004; 135:2106-2111.
179. Tonon T, Harvey D, Qing R et al. Identification of a fatty acid Delta11-desaturase from the microalga Thalassiosira pseudonana. FEBS Lett 2004; 563(1-3):28-34.
180. Tonon T, Sayanova O, Michaelson LV et al. Fatty acid desaturases from the microalga Thalassiosira pseudonana. Febs J 2005; 272(13):3401-12.
181. Wen ZY, Chen F. Heterotrophic production of eicosapentaenoic acid by microalgae. Biotechnol Adv 2003; 21(4):273-94.
182. Lebeau T, Robert JM. Diatom cultivation and biotechnologically relevant products. Part II: Current and putative products. Appl Microbiol Biotechnol 2003; 60(6):624-32.
183. Round FE, Crawford RM, Mann DG. The Diatoms. Cambridge, UK: Cambridge University Press, 1990.
184. Aspinall-O'Dea M, Wentworth M, Pascal A et al. In vitro reconstitution of the activated zeaxanthin state associated with energy dissipation in plants. Proc Natl Acad Sci USA 2002; 99(25):16331-5.
185. Peterson RB, Havir EA. Photosynthetic properties of an Arabidopsis thaliana mutant possessing a defective PsbS gene. Planta 2001; 214:142-152.
186. Holt NE, Zigmantas D, Valkunas L et al. Carotenoid cation formation and the regulation of photosynthetic light harvesting. Science 2005; 307(5708):433-6.
187. Kuroiwa T, Kuroiwa H, Sakai A et al. The division apparatus of plastids and mitochondria. Int Rev Cytol 1998; 181:1-41.
188. Nozaki H, Matsuzaki M, Misumi O et al. Cyanobacterial genes transmitted to the nucleus before divergence of red algae in the Chromista. J Mol Evol 2004; 59(1):103-13.
189. Ciniglia C, Yoon HS, Pollio A et al. Hidden biodiversity of the extremophilic Cyanidiales red algae. Mol Ecol 2004; 13(7):1827-38.
190. Minoda A, Sakagami R, Yagisawa F et al. Improvement of culture conditions and evidence for nuclear transformation by homologous recombination in a red alga, Cyanidioschyzon merolae 10D. Plant Cell Physiol 2004; 45(6):667-71.
191. Ohta N, Matsuzaki M, Misumi O et al. Complete sequence and analysis of the plastid genome of the unicellular red alga Cyanidioschyzon merolae. DNA Res 2003; 10(2):67-77.
192. Kuroiwa T. Mechanism of mitochondrial and plastid divisions: Memory of 3 genome sequences of Cyanidioschyzon merolae as an origin of enkaryote. Tanpakushitsu Kakusan Koso 2005; 50(2):97-110.
193. Miyagishima SY, Nishida K, Mori T et al. A plant-specific dynamin-related protein forms a ring at the chloroplast division site. Plant Cell 2003; 15(3):655-65.
194. Miyagishima SY, Nozaki H, Nishida K et al. Two types of FtsZ proteins in mitochondria and red-lineage chloroplasts: The duplication of FtsZ is implicated in endosymbiosis. J Mol Evol 2004; 58(3):291-303.
195. Nishida K, Misumi O, Yagisawa F et al. Triple immunofluorescent labeling of FtsZ, dynamin, and EF-Tu reveals a loose association between the inner and outer membrane mitochondrial division machinery in the red alga Cyanidioschyzon merolae. J Histochem Cytochem 2004; 52(7):843-9.
196. Miyagishima S, Kuroiwa H, Kuroiwa T. The timing and manner of disassembly of the apparatuses for chloroplast and mitochondrial division in the red alga Cyanidioschyzon merolae. Planta 2001; 212(4):517-28.
197. Miyagishima S, Takahara M, Mori T et al. Plastid division is driven by a complex mechanism that involves differential transition of the bacterial and eukaryotic division rings. Plant Cell 2001; 13(10):2257-68.
198. Miyagishima S, Takahara M, Kuroiwa T. Novel filaments 5 nm in diameter constitute the cytosolic ring of the plastid division apparatus. Plant Cell 2001; 13(3):707-21.
199. Miyagishima S, Itoh R, Aita S et al. Isolation of dividing chloroplasts with intact plastid-dividing rings from a synchronous culture of the unicellular red alga Cyanidioschyzon merolae. Planta 1999; 209(3):371-5.
200. Kuroiwa T. The discovery of the division apparatus of plastids and mitochondria. J Electron Microsc (Tokyo) 2000; 49(1):123-34.
201. Maruyama S, Misumi O, Ishii Y et al. The minimal eukaryotic ribosomal DNA units in the primitive red alga Cyanidioschyzon merolae. DNA Res 2004; 11(2):83-91.
202. Cavalier-Smith T. Membrane heredity and early chloroplast evolution. Trends Plant Sci 2000; 5(4):174-82.

203. Maier UG, Douglas SE, Cavalier-Smith T. The nucleomorph genomes of cryptophytes and chlorarachniophytes. Protist 2000; 151(2):103-9.
204. He Q, Dolganov N, Bjorkman O et al. The high light-inducible polypeptides in Synechocystis PCC6803. Expression and function in high light. J Biol Chem 2001; 276(1):306-314.
205. Wastl J, Sticht H, Maier UG et al. Identification and characterization of a eukaryotically encoded rubredoxin in a cryptomonad alga. FEBS Lett 2000; 471(2-3):191-6.
206. Wastl J, Maier UG. Transport of proteins into cryptomonads complex plastids. J Biol Chem 2000; 275(30):23194-8.
207. Henriksen K, Stipp SLS, Young JR et al. Tailoring calcite: Nanoscale AFM of coccolith biocrystals. Am Mineralogist 2003; 88:2040-2044.
208. Young JR, Davis SA, Bown PR et al. Coccolith ultrastructure and biomineralisation. J Struct Biol 1999; 126(3):195-215.
209. Marsh ME. Regulation of $CaCO_3$ formation in coccolithophores. Comp Biochem Physiol B Biochem Mol Biol 2003; 136(4):743-54.
210. Brown CW, Yoder JA. Coccolithophorid blooms in the global ocean. J Geophys Res 1994; 99:7467-7482.
211. Holligan PM, Groom SB, DSH. What controls the distribution of the coccolithophore, Emiliania huxleyi, in the North Sea? Fish Oceanogr 1993; (2):175-183.
212. Wahlund TM, Hadaegh AR, Clark R et al. Analysis of expressed sequence tags from calcifying cells of marine coccolithophorid (Emiliania huxleyi). Mar Biotechnol (NY) 2004; 6(3):278-90.
213. Wahlund TM, Zhang X, Read BA. Expression profiles from calcifying and noncalcifying cultures of Emiliania huxleyi. J Micropaleontol 2004; 51:145-155.
214. Nguyen B, Bowers RM, Wahlund TM et al. Suppressive subtractive hybridization of and differences in gene expression content of calcifying and noncalcifying cultures of Emiliania huxleyi strain 1516. Appl Environ Microbiol 2005; 71(5):2564-75.
215. Paasche E, Bruback S. Enhanced calcification in the coccolithophorid Emiliania huxleyi (Haptophyceae) under phosphorus limitation. Phycologia 1994; 33:324-330.
216. van Bleijswijk JD, Velduis MJW. In situ gross growth rates of Emiliania huxleyi in enclosures with different phosphate loadings revealed by diel changes in DNA content. Mar Ecol Prog Ser 1995; 121:271-277.
217. Bachvaroff TR, Concepcion GT, Rogers CR et al. Dinoflagellate expressed sequence tag data indicate massive transfer of chloroplast genes to the nuclear genome. Protist 2004; 155(1):65-78.
218. Hackett JD, Yoon HS, Soares MB et al. Migration of the plastid genome to the nucleus in a peridinin dinoflagellate. Curr Biol 2004; 14(3):213-8.
219. Hackett JD, Scheetz TE, Yoon HS et al. Insights into a dinoflagellate genome through expressed sequence tag analysis. BMC Genomics 2005; 6(1):80.
220. Nikaido I, Asamizu E, Nakajima M et al. Generation of 10,154 expressed sequence tags from a leafy gametophyte of a marine red alga, Porphyra yezoensis. DNA Res 2000; 7(3):223-7.
221. Scala S, Carels N, Falciatore A et al. Genome properties of the diatom Phaeodactylum tricornutum. Plant Physiol 2002; 129(3):993-1002.
222. Crepineau F, Roscoe T, Kaas R et al. Characterisation of complementary DNAs from the expressed sequence tag analysis of life cycle stages of Laminaria digitata (Phaeophyceae). Plant Mol Biol 2000; 43(4):503-13.
223. Bergh O, Borsheim KY, Bratbak G et al. High abundance of viruses found in aquatic environments. Nature 1989; 340(6233):467-8.
224. Proctor LM, Fuhrman JA. Viral mortality of marine bacteria and cyanobacteria. Nature 1990; 343:60-62.
225. Culley AI, Lang AS, Suttle CA. High diversity of unknown picorna-like viruses in the sea. Nature 2003; 424(6952):1054-7.
226. Cochlan WP, Wilkner J, Stewart GF et al. Spatial distribution of viruses, bacteria and chlorophyll a in neritic, oceanic and estuarine environments. Mar Ecol Prog Ser 1993; 92:77-87.
227. Paul JH, Rose JB, Jiang SC et al. Distribution of viral abundance in the reef environment of Key Largo, Florida. Appl Environ Microbiol 1993; 59:718-724.
228. Frank H, Moebius K. An electron microscopic study of bacteriophages from marine waters. Helgolander Meeresunters 1987; 41:385-414.
229. Suttle C. Crystal ball. The viriosphere: The greatest biological diversity on Earth and driver of global processes. Environ Microbiol 2005; 7(4):481-2.
230. Mann NH. Phages of the marine cyanobacterial picophytoplankton. FEMS Microbiol Rev 2003; 27(1):17-34.
231. Millard A, Clokie MR, Shub DA et al. Genetic organization of the psbAD region in phages infecting marine Synechococcus strains. Proc Natl Acad Sci USA 2004; 101(30):11007-12.

232. Lindell D, Sullivan MB, Johnson ZI et al. Transfer of photosynthesis genes to and from Prochlorococcus viruses. Proc Natl Acad Sci USA 2004; 101(30):11013-8.
233. Bailey S, Clokie MR, Millard A et al. Cyanophage infection and photoinhibition in marine cyanobacteria. Res Microbiol 2004; 155(9):720-5.
234. Zeidner G, Bielawski JP, Shmoish M et al. Potential photosynthesis gene recombination between Prochlorococcus and Synechococcus via viral intermediates. Environ Microbiol 2005; 7(10):1505-13.
235. Lindell D, Jaffe JD, Johnson ZI et al. Photosynthesis genes in marine viruses yield proteins during host infection. Nature 2005; 438(7064):86-9.
236. Moebius K, Nattkemper H. Bacteriophage sensitivity patterns among bacteria isolated from marine waters. Helgolander Meeresunters 1981; 34:375-385.
237. Suttle CA. The ecological, evolutionary and geochemical consequences of viral infection of cyanobacteria and eukaryotic algae. In: Hurst CJ, ed. Viral Ecology. New York: Academic Press, 2000:248-286.
238. van Etten JL, Graves MV, Muller DG et al. Phycodnaviridae - Large DNA algal viruses. Arch Virol 2002; 147:1479-1516.
239. Delaroque N, Muller DG, Bothe G et al. The complete DNA sequence of the Ectocarpus siliculosus Virus EsV-1 genome. Virology 2001; 287(1):112-32.
240. Lang AS, Culley AI, Suttle CA. Genome sequence and characterization of a virus (HaRNAV) related to picorna-like viruses that infects the marine toxic bloom-forming alga Heterosigma akashiwo. Virology 2004; 320(2):206-17.
241. Brussaard CP, Noordeloos AA, Sandaa RA et al. Discovery of a dsRNA virus infecting the marine photosynthetic protist Micromonas pusilla. Virology 2004; 319(2):280-91.
242. Nagasaki K, Tomaru Y, Takao Y et al. Previously unknown virus infects marine diatom. Appl Environ Microbiol 2005; 71:3528-3535.
243. Shackelton LA, Holmes EC. The evolution of large DNA viruses: Combining genomic information of viruses and their hosts. Trends Microbiol 2004; 12(10):458-65.
244. Iyer LM, Aravind L, Koonin EV. Common origin of four diverse families of large eukaryotic DNA viruses. J Virol 2001; 75(23):11720-34.
245. Sandaa RA, Heldal M, Castberg T et al. Isolation and characterization of two viruses with large genome size infecting Chrysochromulina ericina (Prymnesiophyceae) and Pyramimonas orientalis (Prasinophyceae). Virology 2001; 290(2):272-80.
246. Jacobsen A, Bratbak G, Heldal M. Isolation and characterization of a virus infecting Phaeocystis pouchetii (Prymnesiophyseae). J Phycol 1996; 32:923-927.
247. Van Etten JL, Meints RH. Giant viruses infecting algae. Annu Rev Microbiol 1999; 53:447-94.
248. Müller DG, Kapp M, Knippers R. Viruses in marine brown algae. Adv Virus Res 1998; 50:49-67.
249. Chen F, Suttle CA, Short SM. Genetic diversity in marine algal virus communities as revealed by sequence analysis of DNA polymerase genes. Appl Environ Microbiol 1996; 62:2869-2874.
250. Short SM, Suttle CA. Sequence analysis of marine virus communities reveals that groups of related algal viruses are widely distributed in nature. Appl Environ Microbiol 2002; 68(3):1290-6.
251. Breitbart M, Rohwer F. Here a virus, there a virus, everywhere the same virus? Trends Microbiol 2005; 13(6):278-84.
252. Wilson WH, Schroeder DC, Allen MJ et al. Complete genome sequence and lytic phase transcription profile of a Coccolithovirus. Science 2005; 309(5737):1090-2.
253. Allen MJ, Schroeder DC, Wilson WH. Preliminary characterisation of repeat families in the genome of EhV-86, a giant algal virus that infects the marine microalga Emiliania huxleyi. Arch Virol 2005.
254. Allen MJ, Schroeder DC, Holden MT et al. Evolutionary history of the Coccolithoviridae. Mol Biol Evol 2006; 23(1):86-92.

Chapter 7

Insertional Mutagenesis as a Tool to Study Genes/Functions in *Chlamydomonas*

Aurora Galván,* David González-Ballester and Emilio Fernández

Abstract

The unicellular alga *Chlamydomonas reinhardtii* has emerged during the last decades as a model system to understand gene functions, many of them shared by bacteria, fungi, plants, animals and humans. A powerful resource for the research community is the availability of complete collections of stable mutants for studying whole genome function. In the meantime other strategies might be developed; insertional mutagenesis has become currently the best strategy to disrupt and tag nuclear genes in *Chlamydomonas* allowing forward and reverse genetic approaches. Here, we outline the mutagenesis technique stressing the idea of generating databases for ordered mutant libraries, and also of improving efficient methods for reverse genetics to identify mutants defective in a particular gene.

Chlamydomonas as a Model for Translational Biology

The complexity of biological systems requires the utilization of model organisms to understand physiological and non-physiological (pathological) processes. *Chlamydomonas* has the goodness for such model organism.[1,2] It is easy to grow in the lab under different environmental and nutritional conditions, it has a short generation time (six hours), it is haploid and can also form stable diploids, useful in complementation tests, its genome sequence is mostly known and its preliminary analysis supports the existence of at least 16,000 genes, an important collection of EST (over 300,000) from many nutritional conditions is available, molecular map markers covering the 17 linkage groups of *Chlamydomonas* have been developed[3,4] (www.chlamy.org) and transformation of its three genomes, nuclear, chloroplast and mitochondrial is possible.[5-8] In addition, the easiness to obtain stable mutants by many different strategies offers an invaluable tool for functional genomics. The properties of this organism simplifies physiological and biochemical characterization of isolated mutants. An important mutant collection is available at the *Chlamydomonas* center (www.chlamy.org) as well as several labs have focused their attention on the construction of whole genome mutant libraries.[9,10]

Contributions and perspectives for a translational biology from *Chlamydomonas* to other systems come from important insights in processes such as photosynthesis,[11-13] molecular biology of the flagellar apparatus,[14-16] mineral nutrition,[1,17,18] light perception and circadian rhythm,[19,20] gametogenesis and fertilization.[1,2,16,21] Historically, *Chlamydomonas* has been used for more than 40 years as a key model for studying of photosynthesis and molecular biology of the chloroplast. *Chlamydomonas* has made it possible to isolate the first photosynthetic

*Corresponding Author: Aurora Galván—Departamento de Bioquímica y Biología Molecular, Facultad de Ciencias, Universidad de Córdoba. Campus de Rabanales, Edificio Severo Ochoa, 14071-Córdoba, Spain. Email: bb1gacea@uco.es

Transgenic Microalgae as Green Cell Factories, edited by Rosa León, Aurora Galván and Emilio Fernández.

mutants and the reason for that is the possibility to grow either photoautotrophically in minimal media and light or heterotrophically in acetate and dark. Currently a wide number of genes related to photosynthesis have been analyzed.[1,22]

Concerning mineral nutrition, *Chlamydomonas* has excellent advantages as a plant model to understand processes for acquisition and assimilation of macronutrients (nitrogen, sulfur and phosphorus) as well as micronutrients.[17,23] These and other metabolic routes are well documented and the biochemical tools are also available.[22] In spite of some previous approaches, the regulatory networks connections between different metabolic processes are starting to be weakly developed, as it is the case of phosphorous/sulfur and nitrogen/carbon relations.[24-29] Anyway, additional work and new Systems Biology approaches will have to be developed. As many other organisms, *Chlamydomonas* cells have developed such regulatory mechanisms to adapt to changeable external conditions, but its unicellular nature makes *Chlamydomonas* a first choice system to be used in many studies.

To illustrate the goodness of *Chlamydomonas* in this metabolic area some examples come from nitrogen assimilation where nitrate/nitrite transporters corresponding to two families, *Nrt2* and *Nar1*, were first identified in photosynthetic eukaryotes. *Chlamydomonas* research has contributed to the identification of high-affinity nitrate transporters (HANT), encoded by *Nrt2* genes (*Nrt2.1* and *Nrt2.2*).[30,31] These systems I and II required a second component named *Nar2* in *Chlamydomonas*[31-33] which later on has been identified in the plants barley[34] and *Arabidopsis*.[35] *Chlamydomonas* also contributed to the first identification of the nitrite transporter to the chloroplast encoded by *Nar1.1*.[27] *Nar1* belongs to the FNT transporters (formate nitrite transporter) likely restricted to protist organisms but absent in plants.[28] The complexity of the *Chlamydomonas Nar1* gene family, formed by six genes, is fascinating since the other protists contain just one to three genes. Functionality of NAR1 proteins seems to be related to the transport of nitrite and bicarbonate with predicted plastidic or plasma membrane localizations.[28] Understanding how a network of so many transporters can be efficiently used by single cells is a challenge for which *Chlamydomonas* is an indisputable model.

Chlamydomonas extends further beyond a model for photosynthetis or mineral nutrition, so it is contributing importantly to the knowledge of cilia and flagella functions. Cilia and flagella are ubiquitous organella from protists to humans and contribute to an interesting research area as that on cilia-related human diseases from cystic kidney to obesity, hypertension, and diabetes.[36,37] The *Chlamydomonas* flagellar proteosome revealed that almost 52% of the proteins are conserved in humans.[38]

Finally, *Chlamydomonas* is capable of performing a complex fermentative metabolism, normally more related to prokaryotes and fungi rather than to photosynthetic eukaryotes. The fermentative pattern includes the products formate, ethanol, acetate, glycerol, lactate, carbon dioxide and molecular hydrogen among others.[39,40] This unusual feature is being used for many researchers to use this alga in biotechnological applications (see Chapter 10).[41]

Mutants as a Tool for Functional Genomics

Functional genomics intends to study the relationship existing between a gene sequence and its function to proceed from individual genes to every gene participating in biological processes in the context of a whole cell. Those studies include the characterization of gene expression and that of the encoded protein and its activity. Two major approaches can be used in functional genomics; one is based on the -omics sciences and the other on blockage of gene function. Both strategies are complementary and jointly constitute a powerful tool to understand gene functions. The -omics sciences being used in this algal system provided a valuable amount of information. DNA arrays have been performed to bring information on transcriptome data in response to light, circadian clock, CO_2, gametogenesis program, sulfur and phosphate deprivation.[24-26,29,42-44] Hundreds of genes can change their expression levels when a single condition is altered, but also these changes vary with the time. So, sometimes it is difficult to understand gene functions only on the basis of the gene expression pattern. On the other hand, interfering with the function of a particular gene has demonstrated to be a powerful tool for understanding its function. Several methodologies are available for this interference that include (i) classic mutagenesis, (ii) interference RNA (RNAi), (iii) gene disruption by homologous recombination, and (iv) insertional mutagenesis by non-homologous recombination (including both direct and reverse genetics selections).

Classical Mutagenesis

Classical mutagenesis procedures based on chemical (N-Nitrosoguanidine, ethylmethane sulphonate, etc...) or physical (UV radiation) agents might be extremely disruptive, and thus mutagenesis might thus produce different lesions that have pleiotropic effects. Most classical methods of mutagenesis use to produce frameshift or missense lesions that are generally difficult to locate within a particular gene. For example, it was found that chlorate, a nitrate analogue, commonly used to select nitrate assimilation mutants causes general chemical mutagenesis in DNA, probably due to oxidation, and these mutations might cause unknown phenotype effects difficult to control.[45] Notwithstanding, the isolation of mutants has been a powerful tool to understand the different biological phenomena studied in *Chlamydomonas* and its mapping relative to each other, together with RFLPs with another *Chlamydomonas* strain, allowed a first good definition of the genetic map of the alga (http://www.chlamy.org/mapkit/markers.html).[46] This map has been enhanced with the identification of molecular markers so that 264 of them, including sequence-tagged sites (STS), were mapped to the 17 known linkage groups of *Chlamydomonas* covering approximately 1,000 centimorgans (cM). This detailed genetic map greatly facilitates isolation of genes of interest by using positional cloning methods.[3] Also, the availability of a PCR-based marker kit, that use bulked segregant analysis and marker duplexing, allows screening the genome at 20-30 cM (http://www.chlamy.org/kit.html).[47] Once the mutation is mapped, and if linked to a marker, transformation with available BACs, plasmids, or cosmids DNA could be used to recover a wild-type phenotype and so to identify the particular gene.

RNAi

One of the interesting alternatives that have been successfully employed for the direct isolation of the desired mutants of a particular gene (reverse genetics) is the antisense mRNA strategy or RNAi (interference RNA).[48,49] This technique is being used in many biological systems from protozoa to metazoans. RNAi is based on conserved cellular mechanisms acting in defense responses against viruses and transposons and in endogenous gene regulation.[50] It is generaly accepted that silencing occurs by a mechanism that involves a ribonuclease III-like enzyme, Dicer, that processes long dsRNAs into approximately 25-nt small interfering RNAs (siRNAs), that in turn are incorporated into the RNA-induced silencing complex (RISC); the siRNAs in this multiprotein complex are used to identify and cleave homologous mRNAs.[48,50,51]

RNAi induced by introduction of exogenously synthesized dsRNAs or siRNAs into cells produce transient and not heritable silencing effects.[48,49] Thus it is advantageous to endogenously express the RNAi from a construction properly integrated in the genome. Efficient post-transcriptional gene silencing (PTGS) in *Chlamydomonas* was achieved by introducing intron-containing gene fragments directly linked to their intron-less antisense counterpart (Fig. 1A).[52] However, the level of gene silencing is often variable, depending on the type of construct, transgene copy number, site of integration, target gene, and integrity of the integrated transgene.[51,52] To overcome these problems, a tandem inverted repeat system for selection of effective transgenic RNAi strains has been described in *Chlamydomonas*.[53] The method consists of tandem inverted repeat (TIR) transgenes that consistently trigger co-silencing of the *Maa7* gene (encoding tryptophan synthase b subunit) with a selectable RNAi-induced phenotype together with another gene of interest (Fig. 1B). Since transgenes in the genome are inherited, this approach could be used to prepare collections of epi-mutants to study gene function.

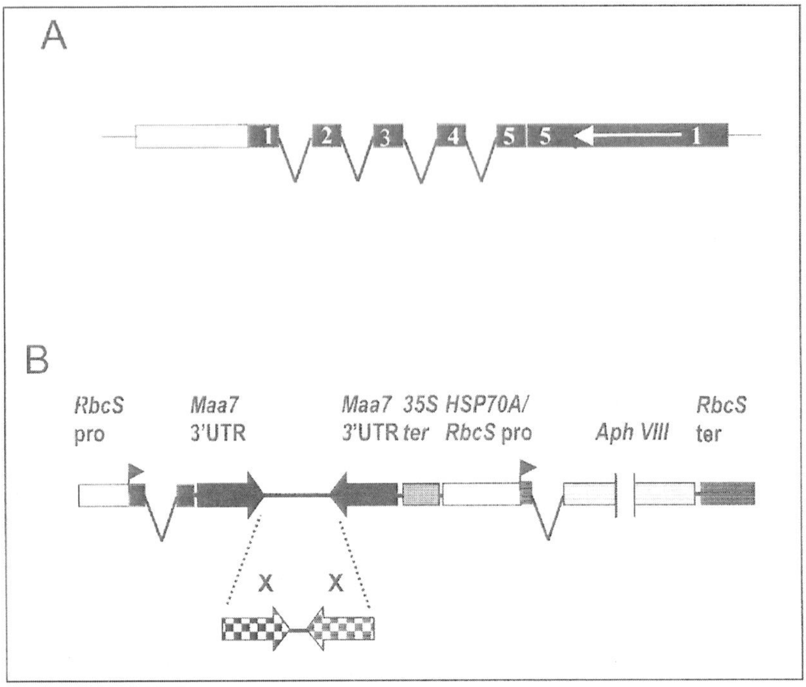

Figure 1. Scheme of constructs for PTGS of a gene of interest. A) The genomic DNA of a fragment of the gene (for example with five exons) is fused with the corresponding cDNA in opposite orientation.[52] B) Cosilencing of tryptophan synthase b subunit (*Maa7*) and another target gene (X) is induced by triggering RNAi production from the expression of tandem *Maa7*/X inverted repeat (IR) transgenes.[53] Promoter and terminating sequences used are schematized.

Several reports have been published about the antisense strategy in *Chlamydomonas*, reviewed critically by Schroda.[51]

Directed Gene Disruption by Homologous Recombination

Gene disruption by homologous recombination could be in a near future an attractive tool for getting knock-out genes in *Chlamydomonas*, but now is still inefficient and substantial efforts are still required to improve it. Nuclear homologous recombination has been observed in *Chlamydomonas* at a very low frequency,[54,55] but recent results have shown that it could be more efficiently achieved by using single-stranded DNA.[56] A previously introduced truncated aminoglycoside 3'-phosphotransferase gene (*AphVIII*) was repaired by homologous recombination with the single-stranded non-defective DNA. By this method, non-homologous DNA integration seems to be reduced more than 100 times in comparison to the use of double-stranded DNA.[56] Thus, this method might be very useful to direct targeting of *Chlamydomonas* nuclear genes.

Insertional Mutagenesis

Insertional mutagenesis has become a popular method to disrupt nuclear genes in *Chlamydomonas* allowing forward genetics approaches based on screening for particular phenotypes and the identification of the tagged genomic region. When *Chlamydomonas* is transformed with an exogenous DNA marker, it is integrated randomly into the nuclear genome and produces from just a single insertion to deletions of 5 to 57 kb.[6,57] This makes it possible to disrupt, tag and clone any non-essential gene, constituting a powerful approach to study many cellular processes. In contrast to chemical mutagenesis, DNA integration events appear to occur at a relatively low frequency, thus avoiding the appearance of multiple mutations in any given strain,[6] and in addition, insertions or deletions are easier to identify than minor classical mutations. Since the first report of this technique to clone flagellar genes in *Chlamydomonas reinhardtii*,[58] it has been extensively used to isolate mutants affected in many different biological processes such as photosynthesys and photoxidative stress,[9] phototaxis,[59] nitrogen metabolism,[10,60] sulfur metabolism,[26,61] carbon metabolism,[62,63] H_2 production,[64] or biology of the cytoplasmic microtubules,[65] etc. Reviews on nuclear transformation in *Chlamydomonas* as well as insertional mutagenesis and its application on forward and reverse genetics have been reported.[6,66]

Insetional mutagenesis process basically consists on the following steps:

1. Cell transformation
2. Selection of transformants
3. Screening for particular phenotypes
4. Identification of the genomic region tagged
5. Recovering of the wild type phenotype with the candidate gene.

Each of these steps require considerations attending to the best choice for strains, selectable marker genes, transformation method, phenotypes to screen and identification of the tagged gene.

Cell Transformation

Among the different techniques used to transform *Chlamydomonas* such as particle bombardment, electroporation or glass beads (see Chapter 1), the last two methods have become the most popular and routine ones for nuclear transformation because of its simplicity, cost and success to any purpose.[67-69] These methods are successfully used for the transformation of cell-wall-deficient mutants (*cw* mutants), or wild type strains after removing the cell wall by treatment with autolysin. In general, *cw* mutants render better efficiency of transformation in relation to that from autolysin treated strains but some *cw* mutants transformed have difficulties in subsequent genetic crosses.

The generation of mutant collections by insertional mutagenesis requires at least two prerequisites to be useful, a high efficiency of transformation and a low number of integrated copies of the DNA marker. The efficiency of transformation is a key point when a whole genome is going to be tagged since many thousands transformants have to be obtained and screened on a reasonable time. According to the *Chlamydomonas* genome size (~10^8 bp) and the deletion range described when a DNA tag is integrated, Pazour and Witman[66] estimated the number of mutant collections to be screened and the probability to find a mutation in a gene of interest. They suggest that 60,000 transformants is a reasonable number to screen for having 95% chances to get a mutation in a particular non-lethal gene. This number it is estimated for a mean deletion of 5 kb, but it could change significantly if other average deletion size is considered. (to 30,000 or 15,000 if deletion sizes are 10 or 20 kb, respectively. On the other hand, reducing the amount of marker DNA employed during the transformations could decrease to a low number the DNA marker copies integrated in the genome that could be as low as one.[9,10,69] A single DNA marker insertion per transformant facilitates the correlation between mutant gene and phenotype, whereas a higher number will decrease the number of transformants needed to cover a mutagenized genome.

Anyway, a real number of mutants to saturate the genome is unpredictable on the basis of the variability of deletions sizes, the number of insertions, and also on a not totally random distribution of the insertions, since hot and cold spots might exist, although this has not been examined in *Chlamydomonas* yet. In *Arabidopsis* over 225,000 T-DNA mutants have been obtained to get saturation of the gene spacing, where 88,000 identified insertions cover almost 74% of genes predicted in the plant.[70] The integration site bias has been analyzed to show a non-uniform chromosomal distribution of integration events and preferent sites (hot spots). The density of the integration sites correlated with gene density in terms that gene-rich regions (chromosomes arms) become hot spots but pericentromeric regions (low gene density, high concentration of pseudogenes and silent transposons) seems to be cold spots.[70] In addition to cold/hot spots, essential genes for cell survival should not be targeted, with exceptions for metabolic processes in which the requirement of the gene can be bypassed by alternative metabolic pathways.

Selection of Transformants and Screening for Phenotypes

Selection of the mutants depends on the gene marker used for mutagenesis, on the recipient *Chlamydomonas* strain. In Table 1 a number of insertional mutant collections and marker used to identify genes from different biological processes are indicated. The selectable gene markers used in *Chlamydomonas* have been previously reviewed[6,41] (see Chapter 1) and could be classified into two types, those that complement auxotrophic or no photosynthetic mutants (*Nia1, Arg7, Nic7, Por1, Oee1, AtpC*) or those that confer resistance against antibiotic (*Ble, AphVIII, Cry, NptII*). The first ones are very effective to complement stable mutants and avoid false positives. However, this strategy is not applicable to wild-type cell and requires a particular mutant. For example, *Nia1* gene marker complements the nitrate reductase mutant and the screening of Nit^+ colonies has been extensively used used to identify genes involved on biogenesis of flagella,[58,66] resistance to salinity,[71] CO_2 signal,[62] ammonium regulation,[60] etc. Having in mind the existence of networking connections between different metabolic or biological processes, special care should be taken in choosing those *Chlamydomonas* strains useful for wide-genome insertional mutant generation and for forward screening analysis. Many of the strains in the *Chlamydomonas* Genetic Center derive from a "wild-type" strain carrying the Nia1, *Nit2* mutations and are thus inappropriate for transformation with *Nia1*.[6,67] Even using drug resistance markers, insertional mutants obtained could be limited for functional genomic studies of some metabolic processes.

Genes for drug resistance like *AadA* from eubacteria which confers resistance to spectinomycin,[72] *AphVIII* from *Streptomyces rimosus* which confers resistance to paromomycin,[73] and *Ble* from *Streptoalloteichus hindustanus* which confers resistance to zeomicyn[74] are usefull markers to obtain *Chlamydomonas* mutants.[9,10,26] These markers have special advantages for a wide-genome insertional mutant strategy, since they are applicable to any strain and their small size and use in a linearized form should facilitate the isolation of the tagged genes. However, other inconveniences have to be considered such as the antibiotic dependent-conditions and the gene silence phenomenon of foreign genes, that for the *AadA* gene results in a

Table 1. Insertional mutagenesis as a tool to identify genes/functions in Chlamydomonas

Goal of the Study	References	Gene or Marker Used	Molecular Identification of Tagged Genes	Conservation of the Collection of Mutants
Flagella and motility	58,66	*Nia1*	-	+
Nitrogen metabolism	10,60,80	*Nia1 AphVIII*	+	-
Sulfur metabolism	26	*AphVIII*	+	-
Carbon metabolism	81	*Arg7*	-	-
	62	*Nia1*	+	-
Photosynthesis	82	*Arg7*	-	-
	9	*Ble*	+	-
Hydrogen photoproduction	64	*Arg7*	+	-

spectinomycin unstable phenotype in 50% of the transformants after subsequent growth under non-selective conditions (see Chapter 1).[72] The resistance to bleomicyn depends on the *Chlamydomonas* strain as well on the culture conditions used,[74] so it is important to adjust the antibiotic concentrations in the selection medium. In addition, the action mechanism of bleomycin which breaks DNA strands[75] may cause mutations independent on the marker gene insertion (Llamas et al, unpublished). In contrast, the paramomycin resistance gene is particularly useful for insertional mutagenesis since the antibiotic is devoid of side effects.[10]

Screening of insertional mutants based on the use of strains that carry a reporter gene under a specific promoter is an invaluable method for isolating regulatory circuit's mutants as shown for nitrate assimilation regulation by using the *Nia1* gene promoter fused to the arylsulfatase (ARS) reporter gene.[10] By exploring the particular conditions under which the promoter of study would be regulated, transformants showing an alteration of the reporter activity would correspond to regulatory mutants affected in genes controlling the activity of the studied promoter. To this purpose it is important to set up transformation conditions so that on average a single copy of the marker DNA is inserted in transformants to facilitate the correlation of the phenotype to the insertion of the marker. Even so, and probably due to the mechanism of action of bleomycin, transformants resistant to this antibiotic can be recovered with mutations at positions unlinked to the transforming DNA.[9]

Identification of the Genomic Region Tagged

A critical issue in any insertional mutagenesis approach is identifying the genomic region adjacent to the marker DNA used for transformation, so that the insertion of the tag can be related to particular gene functions affected. Different methods have been used for this purpose, some of which need a large dosage of time and effort.[58,62] A popular PCR-based method was set up by Liu and coworkers, Thermal Asymmetric Interlaced PCR (TAIL-PCR), and used in different systems for rapid and efficient isolation of DNA segments adjacent to marker DNA.[76,77] The technique is based on the use of specific primers from the ends of known sequences in combination to degenerated primers aligning randomly in the genome. Though this method has been used in *Chlamydomonas*, unspecificity problems and low quantity of the bands obtained complicate the success in organisms with a high-complexity genome.

Another random PCR-based technique was designed to efficiently identify regions flanking a marker DNA in insertional mutants of *Chlamydomonas*.[78] The technique, named Restriction Enzyme Site-Directed Amplification PCR (RESDA-PCR), is based on the random distribution of frequent restriction sites in a genome and on a special design of primers. As shown in Figure 2, degenerated primers include at the 3' end a low degenerated sequence including a restriction site sequence linked by a polyinosine bridge to a specific adapter sequence at the 5' end. This method has probed to be fast, reliable, and can be extrapolated to any organism and marker DNA by designing the appropriate primers.

A third technique has been successfully used to determine the genomic regions that are flanking the DNA marker. It consists in an adaptation of the Easy Gene Walking genome walker technique where the whole genome is cut with an endonuclease which doesn't cut inside the marker and then specific adaptors are ligated to the end of the generated restriction fragments. Specific primers present in the marker and in the adaptor sequences are used to get amplification of the fragments containing the marker.[27,78]

The amplified bands can be directly sequenced and assigned to a particular genomic region by performing a blast with the *Chlamydomonas* genome (http://www.chlamy.org/) that easily allows knowing the size of the deletion and the possible gene(s) responsible of the phenotype.

To know how many genes in a genome region are affected by the insertion it is important to determine the size of the deletions obtained by insertional mutagenesis. Amplification with specific primers annealing to the other end of the marker DNA-that might contain vector sequence- could be performed by the same RESDA-PCR technique provided that no more sequences of the marker DNA annealing to the specific primers are present in the transformed

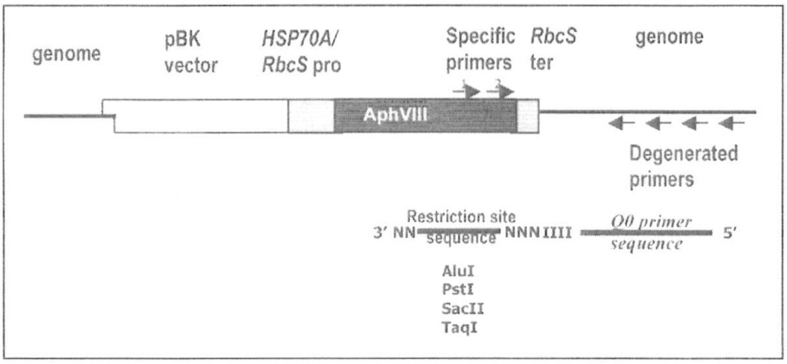

Figure 2. Scheme of the RESDA-PCR technique used to identify sequences adjacent to a DNA marker insertion. Promoter and terminating sequences used are schematized. A first round of amplification is performed combining this degenerated primers with a primer (1) aligning to the marker DNA, that is followed by a second round of nested amplification with two specific primers one to the marker DNA (2) and another to the degenerated primers (Q0).[78]

strain. Clean transformations should be performed by using DNA with only the selectable marker gene.

Recovery of the Wild Type Phenotype with the Candidate Gene

That insertion of the marker DNA is the responsible for causing the selected phenotype can be first shown by performing genetic analysis. These genetic crosses are important to demonstrate a cosegregation between a particular phenotype and the inserted marker DNA and then to put into value the subsequent cloning of the DNA flanking the insertion.

From the analysis of the genomic region where the selectable maker has inserted the candidate gene or genes responsible for the mutant phenotype can be estimated, since deletions may remove more than one gene. Then, genomic DNA for these candidate genes can be obtained from the BAC clone identified in the BAC collection (http://www.chlamy.org/bac_details.html), and by transformation with the proper gene the wild type phenotype would be recovered.

An ordered, insertionally tagged mutant collection was constructed in *Chlamydomonas* by using the paramomycin-resistance gene *AphVIII* in a strain carrying the arylsulfatase gene under control of the nitrate reductase promoter. The mutants contained mostly a single copy of the marker and the screenings showed nitrate utilization deficiencies and ammonium insensitive, overexpressing, or nitrate insensitive phenotypes. This collection has allowed a first approach of functional genomics to identify genes involved in the regulatory circuits of nitrate assimilation. By forward genetics screenings mutants were identified and sequences flanking insertion sites obtained, facilitating the identification of putative genes responsible for specific phenotypes.[10] Another mutant collection was constructed with plasmids conferring zeocin resistance, and insertional mutants were selected in the dark on acetate-containing medium to recover light sensitive and non-photosynthetic mutants. The phenotypic screenings allowed identifying photosynthesis-related mutants, their molecular analysis showed that each mutant contains an average of 1.4 insertions, and genetic analyses indicated that only about 50% of the mutations are tagged by the transforming DNA.[9] This whole mutant collection has not been preserved for further use.

Reverse Genetics

The *Chlamydomonas* genome project is an invaluable tool for functional genomic approaches. To this end a preserved and ordered insertional mutant library is important for isolating mutants of interest. This can be performed either by a direct phenotype screening to identify its responsible gene (forward genetics) or by searching for a mutant with the gene of interest tagged for learning from its phenotype (reverse genetics). Only one mutant library has been obtained and preserved in liquid nitrogen with these purposes[10] (Table 1) that might also be useful to identify non-essential genes related to any cell or metabolic process.

In contrast to other systems having an efficient homologous recombination where reverse genetics can be performed by directed gene disruption, in *Chlamydomonas* homologous recombination efficiency is rather small (see above in this chapter).[54] Reverse genetics has been addressed in *Chlamydomonas* by two methods. One is based on the fact that integration of the selectable marker generates RFLPs and/or deletions that can be detected by DNA hybridization,[66] and the other by PCR approaches.[79]

Pazour and Wittman[66] have described in detail a hybridization method for reverse genetics that consists on: (a) DNA is purified from each of the strains, (b) the DNA is analyzed by Southern hybridization using the gene of interest as the probe, (c) among most strains showing the wild-type (WT) pattern of hybridization, insertional mutants show an RFLP or deletion, (d). The phenotypes of the identified mutants are then characterized to determine the role of that gene. Since the number of mutants to be screened by DNA hybridization might be too high, it is advised to prescreen the collection to reduce this number. For example, if the gene is thought to function in the flagella, it is advised to isolate a collection of mutations that affect motility. This method has been used with success for identifying flagella motility mutants affected in some *oda* and *fla* genes However, to perform this preselection is not possible for those mutants with subtle, no apparent, or unknown phenotype beforehand and thus the hybridization procedure for thousands of strains can become prohibitive in costs and amount of work without guarantee that the searched mutant has been obtained.

The method for reverse genetics in *Chlamydomonas* by using the PCR screening relies on the ordered insertional mutant library.[10] Since the number of mutant strains to analyze is very high, about 22,000, they have been distributed in 222 groups of 96 selected insertional mutants that are grown together. DNA is extracted for each of these groups. Twenty three ordered mixtures of 10 "supergroups" are obtained, each representing 960 mutants. PCR with specific primers from the gene of interest is then performed so that 23 PCR reactions per primer might be enough to screen the whole mutant collection (Fig. 3). When bands are amplified from a particular supergroup, PCRs on each of the groups for this supergroup is performed to know

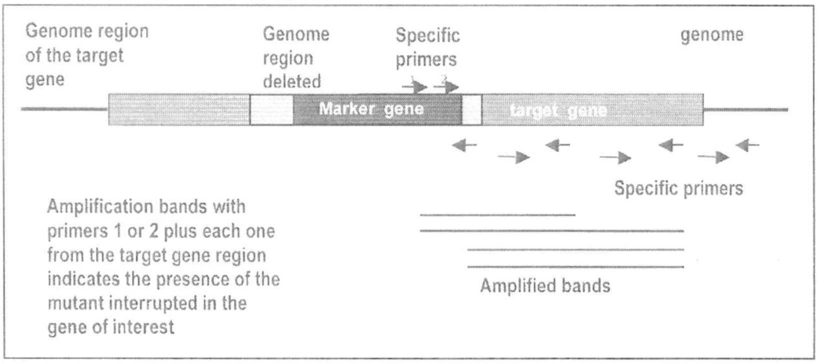

Figure 3. Scheme of the PCR method for screening of mutants by reverse genetics. A target gene of interest and an insertion of the marker gene are represented in an insertional mutant. By PCR, it would be possible to amplify specific bands by using each of the primers in the marker gene and several of the primers in the target gene in adequate combinations. Only those primers with an appropriate distance and orientation would give a productive amplification. The sequencing of bands obtained would show both specificity of the band and the position where the maker DNA has been inserted.[79]

which one contains the mutant. Identified the group, PCR from colonies grown from a plate for this group is carried out to get the insertional mutant in the desired gene.[79]

The mutant can then be crossed to a wild-type cell line and cosegregation of the mutant phenotype with that of the selectable marker shows that the appropriate mutant was selected, so that transformation with the wild type gene would be possible. Since no predictable phenotype can be obtained for a mutant in a particular gene, molecular and physiological characterization of this strain will allow knowing what phenotype could be expected for it and from its reversion to the wild type phenotype upon transformation with the corresponding gene.

Future Perspectives

The development of insertional mutagenesis has greatly simplified the generation of new mutations in particular genes for which a screenable phenotype could be detected, thus facilitating the subsequent cloning of the corresponding gene. The constructions of mutant libraries that are preserved will be enormously useful for the *Chlamydomonas* community achievements that can be boosted if marker gene tags such as T-DNA are developed. Furthermore, ordered insertional mutant libraries make reverse genetics approaches possible, encouraging the understanding of functions for the thousands of new genes identified in the genome project.

References

1. Harris EH. Chlamydomonas as a model organism. Annu Rev Plant Physiol Plant Mol Biol 2001; 52:363-406.
2. Gutman BL, Nigyogi KK. Chlamydomonas and Arabidopsis: A dynamic duo. Plant Physiol 2004; 135:607-610.
3. Kathir P, LaVoie M, Brazelton WJ et al. Molecular map of the Chlamydomonas reinhardtii nuclear genome. Eukaryot Cell 2003; 2:362-379.
4. Li JB, Lin SP, Jia HG et al. Analysis of Chlamydomonas reinhardtii genome structure using large-scale sequencing of regions on linkage groups I and III. J Eukaryotic Microbiology 2003; 50:145-155.
5. Boyton JE, Gillham NW, Harris EH et al. Chloroplast transformation in Chlamydomonas with high velocity microprojectiles. Science 1988; 240:1534-1538.
6. Kindle KL. Nuclear transformation: Technology and applications. In: Rochaix JD et al, eds. The Molecular Biology of Chloroplasts and Mitochondria in Chlamydomonas. Kluwer Academic Publishers, 1998:41-61.
7. Randolph-Anderson BL, Boynton JE, Gillham NW et al. Further characterization of the respiratory deficient dum-1 mutation of Chlamydomonas reinhardtii and its use as a recipient for mitochondrial transformation. Mol Gen Genet 1993; 236:235-244.
8. Remacle C, Cardol P, Coosemans N et al. High-efficiency biolistic transformation of Chlamydomonas mitochondria can be used to insert mutations in complex I genes. Proc Natl Acad Sci USA 2006; 103:4771-4776.
9. Dent R, Haglund C, Chin B et al. Functional genomics of eukaryotic photosynthesis using insertional mutagenesis of Chlamydomonas reinhardtii. Plant Physiol 2005; 137:545-556.
10. González-Ballester D, de Montaigu A, Higuera JJ et al. Functional genomics of the regulation of the nitrate assimilation pathway in Chlamydomonas. Plant Physiol 2005; 137:522-533.
11. Harris EH. The Chlamydomonas Sourcebook. New York: Academic Press, 1989.
12. Rochaix JD. Chlamydomonas reinhardtii as the photosystthetic yeast. Annu Rev Genet 1995; 29:209-230.
13. Rochaix JD. Chlamydomonas, a model system for studying the assembly and dynamics of photosynthetic complexes. FEBS Lett 2002; 529:34-38.
14. Dutcher SK. Flagellar assembly in two hundred and fifty easy-to-follow steps. Trends Genet 1995; 11:398-404.
15. Rosembaum JL, Cole DG, Diener DR. Intraflagellar transport; the eyes have it. J Cell Biol 1999; 144:385-388.
16. Silflow C, Lefebvre PA. Assembly and motility of eukaryotic cilia and flagella: Lessons from Chlamydomonas reinhardtii. Plant Physiol 2001; 127:1500-1507.
17. Grossman A, Takahashi H. Macronutrient utilization by photosynthetic eukaryotes and the fabric of interactions. Annu Rev Plant Physiol Plant Mol Biol 2001; 52:163-210.
18. Galván A, Fernández E. Eukaryotic nitrate and nitrite transport. Cell Mol Life Sci 2001; 58:225-233.

19. Mittag M, Wagner V. The circadian clock of the unicellular eukaryotic model organism Chlamydomonas reinhardtii. Biol Chem 2003; 384:689-695.
20. Breton G, Kay SA. Circadian rhythms lit up in Chlamydomonas. Genome Biol 2006; 7:215.
21. Ferris PJ, Goodenough UW. Mating type in Chlamydomonas is specified by mid, the minus-dominance gene. Genetics 1997; 146:859-869.
22. In: Rochaix JD, Goldschmidt-Clermont M, Merchant S, eds. The Molecular Biology of Chloroplasts and Mitochondria in Chlamydomonas. Dordrecht: Kluwer, 1998.
23. Fernández E, Galván A, Quesada A. Nitrogen assimilation and its regulation. In: Rochaix JD, Goldschmidt-Clermont M, Merchant S, eds. The Molecular Biology of Chloroplasts and Mitochondria in Chlamydomonas. Dordrecht: Kluwer, 1998:637-659.
24. Moseley JL, Chang CW, Grossman AR. Genome-based approaches to understanding phosphorus deprivation responses and PSR1 control in Chlamydomonas reinhardtii. Eukaryot Cell 2006; 5:26-44.
25. Zhang Z, Shrager J, Jain M et al. Insights into the survival of Chlamydomonas reinhardtii during sulfur starvation based on microarray analysis of gene expression. Eukaryotic Cell 2004; 3:1331-1348.
26. Pollock SV, Pootakham W, Shibagaki N et al. Insights into the acclimation of Chlamydomonas reinhardtii to sulfur deprivation. Photosynth Res 2005; 86:475-489.
27. Rexach J, Fernández E, Galván A. The Chlamydomonas reinhardtii Nar1 gene encodes a chloroplast membrane protein involved in nitrite transport. Plant Cell 2000; 12:1441-1453.
28. Mariscal V, Moulin P, Orsel M et al. Differential Regulation of the Chlamydomonas Nar1 gene family by carbon and nitrogen. Protist 2006; 157:421-433.
29. Miura K, Yamano T, Yoshioka S et al. Expression profiling-based identification of CO_2-responsive genes regulated by CCM1 controlling a carbon-concentrating mechanism in Chlamydomonas reinhardtii. Plant Physiol 2004; 135:1595-1607.
30. Quesada A, Galván A, Schnell R et al. Five nitrate assimilation related loci are clustered in Chlamydomonas reinhardtii. Mol Gen Genet 1993; 240:387-394.
31. Quesada A, Galván A, Fernández E. Identification of nitrate transporter genes in Chlamydomonas reinhardtii. Plant J 1994; 5:407-419.
32. Galván A, Quesada A, Fernández E. Nitrate and nitrite are transported by different specific transport systems and by a bispecific transporter in Chlamydomonas reinhardtii. J Biol Chem 1996; 271:2088-2092.
33. Zhou JJ, Fernández E, Galván A et al. A high affinity nitrate transport system from Chlamydomonas requires two gene products. FEBS Lett 2000; 466:225-227.
34. Tong Y, Zhou JJ, Li ZS et al. A two-component high-affinity nitrate uptake system in barley. Plant J 2005; 41:442-450.
35. Okamoto M, Kumar A, Li W et al. High-affinity nitrate transport in roots of Arabidopsis depends on expression of the NAR2-like gene AtNRT3.1. Plant Physiol 2006; 140:1036-1046.
36. Zein LE, Omran H, Bouvagnet P. Lateralization defects and ciliary dyskinesia: Lessons from algae. Trends Genet 2003; 19:162-167.
37. Snell WJ, Pan J, Wang Q. Cilia and flagella revealed: From flagellar assembly in Chlamydomonas to human obesity disorders. Cell 2004; 117:693-697.
38. Pazour G, Agrin N, Leszyk J et al. Proteomic analysis of a eukaryotic cilium. J Cell Biol 2005; 170:103-113.
39. Gfeller RP, Gibbs M. Fermentative metabolism of Chlamydomonas reinhardtii: I. Analysis of fermentative products from starch in dark and light. Plant Physiol 1984; 75:212-218.
40. Hemschemeier A, Happe T. The exceptional photofermentative hydrogen metabolism of the green alga Chlamydomonas reinhardtii. Biochem Soc Trans 2005; 33:39-41.
41. León-Bañares R, González-Ballester D, Galván A et al. Transgenic microalgae as green cell-factories. Trends Biotechnol 2004; 22:45-52.
42. Im CS, Zhang Z, Shrager J et al. Analysis of light and CO_2 regulation in Chlamydomonas reinhardtii using genome-wide approaches. Photosynth Res 2003; 75:111-125.
43. Kucho K, Okamoto K, Tabata S et al. Identification of novel clock-controlled genes by cDNA macroarray analysis in Chlamydomonas reinhardtii. Plant Mol Biol 2005; 57:889-906.
44. Abe J, Kubo T, Takagi Y et al. The transcriptional program of synchronous gametogenesis in Chlamydomonas reinhardtii. Curr Genet 2004; 46:304-315.
45. Prieto R, Fernández E. Toxicity of and mutagenesis by chlorate are independent of nitrate reductase activity in Chlamydomonas reinhardtii. Mol Gen Genet 1993; 237:429-438.
46. Ranum LPW, Thompson MD, Schloss JA et al. Mapping flagellar genes in Chlamydomonas using restriction fragment length polymorphisms. Genetics 1988; 120:109-122.
47. Rymarquis LA, Handley JM, Thomas M et al. Beyond complementation: Map-based cloning in Chlamydomonas reinhardtii. Plant Physiol 2005; 137:557-566.

48. Dykxhoorn DM, Novina CD, Sharp PA. Killing the messenger: Short RNAs that silence gene expression. Nat Rev Mol Cell Biol 2003; 4:457-467.
49. Waterhouse PM, Helliwell CA. Exploring plant genomes by RNA-induced gene silencing. Nat Rev Genet 2003; 4:29-38.
50. Cerutti H. RNA interference: Traveling in the cell and gaining functions? Trends Genet 2003; 19:39-46.
51. Schroda M. RNA silencing in Chlamydomonas: Mechanisms and tools. Curr Genet 2006; 49:69-84.
52. Fuhrmann M, Stahlberg A, Govorunova E et al. The abundant retinal protein of the Chlamydomonas eye is not the photoreceptor for phototaxis and photophobic responses. J Cell Sci 2001; 114:3857-3863.
53. Rohr J, Sarkar N, Balenger S et al. Tandem inverted repeat system for selection of effective transgenic RNAi strains in Chlamydomonas. Plant J 2004; 40:611-621.
54. Nelson JA, Lefebvre PA. Targeted disruption of the NIT8 gene in Chlamydomonas reinhardtii. Mol Cell Biol 1995; 15:5762-5769.
55. Sodeinde OA, Kindle KL. Homologous recombination in the nuclear genome of Chlamydomonas reinhardtii. Proc Natl Acad Sci USA 1993; 90:9199-9203.
56. Zorin B, Hegemann P, Sizova I. Nuclear-gene targeting by using single-stranded DNA avoids illegitimate DNA integration in Chlamydomonas reinhardtii. Eukaryotic Cell 2005; 4:1264-1272.
57. Cenkci B, Petersen JL, Small GD. REX1, a novel gene required for DNA repair. J Biol Chem 2003; 278:22574-22577.
58. Tam LW, Lefebvre PA. Cloning of flagellar in Chlamydomonas reinhardtii by DNA insertional mutagenesis. Genetics 1993; 135:375-384.
59. Pazour GJ, Sineshchekov OA, Witman GB. Mutational analysis of the phototransduction pathway of Chlamydomonas reinhardtii. J Cell Biol 1995; 131:427-440.
60. Prieto R, Dubus A, Galván A et al. Isolation and characterization of two regulatory mutants for nitrate assimilation in Chlamydomonas reinhardtii. Mol Genet Genom 1996; 251:461-471.
61. Davies JP, Yildiz FH, Grossman A. Sac1, a putative regulator that is critical for survival of Chlamydomonas reinhardtii during sulfur deprivation. EMBO J 1996; 15:2150-2159.
62. Yoshioka S, Taniguchi F, Miura K et al. The novel Myb transcription factor LCR1 regulates the CO_2-responsive gene Cah1, encoding a periplasmic carbonic anhydrase in Chlamydomonas reinhardtii. Plant Cell 2004; 16:1466-1477.
63. Zabawinski C, van den Koornhuyse N, d'Hulst N et al. Starchless mutants of Chlamydomonas reinhardtii lack the small subunit of a heterotetrameric ADP-glucose pyrophosphorylase. J Bacteriol 2001; 183:1069-1077.
64. Posewitz MC, Smolinski SL, Kanakagiri S et al. Hydrogen photoproduction is attenuated by disruption of an isoamylase gene in Chlamydomonas reinhardtii. Plant Cell 2004; 16:2151-2163.
65. Horst CJ, Fishkind DJ, Pazour GJ et al. An insertional mutant of Chlamydomonas reinhardtii with defective microtubule positioning. Cell Motil Cytoskeleton 1999; 44:143-154.
66. Pazour GJ, Witman GB. Forward and reverse genetic analysis of microtubule motors in Chlamydomonas. Methods 2000; 22:285-298.
67. Kindle KL, Schnell RA, Fernández E et al. Stable nuclear transformation of Chlamydomonas using the Chlamydomonas gene for nitrate reductase. J Cell Biol 1989; 109:2589-2601.
68. Kindle KL. High-frequency nuclear transformation of Chlamydomonas reinhardtii. Proc Natl Acad Sci USA 1990; 87:1228-1232.
69. Shimogawara K, Fujiwara S, Grossman A et al. High-efficiency transformation of Chlamydomonas reinhardtii by electroporation. Genetics 1998; 148:1821-1828.
70. Alonso JM, Stepanova AN, Leisse TH et al. Genome-wide insertional mutagenesis of Arabidopsis thaliana. Science 2003; 301:653-657.
71. Prieto R, Pardo JM, Niu X et al. Salt-sensitive mutants of Chlamydomonas reinhardtii isolated after insertional tagging. Plant Physiol 1996; 112:99-104.
72. Cerutti H, Johnson AM, Gillham NW et al. A eubacterial gene conferring spectinomycin resistance on Chlamydomonas reinhardtii: Integration into the nuclear genome and gene expression. Genetics 1997; 145:97-110.
73. Sizova I, Fuhrmann M, Hegemann P. A Streptomyces rimosus aphVIII gene coding for a new type phosphotransferase provides stable antibiotic resistance to Chlamydomonas reinhardtii. Gene 2001; 277:221-229.
74. Hecht SM. Bleomycin: New perspectives on the mechanism of action. J Nat Prod 2000; 63:158-168.
74. Stevens DR, Rochaix JD, Purton S. The bacterial phleomycin resistance gene ble as a dominant selectable marker in Chlamydomonas. Mol Gen Genet 1996; 251:23-30.

76. Liu YG, Whittier RF. Thermal asymmetric interlaced PCR: Automatable amplification and sequencing of insert end fragments from P1 and YAC clones for chromosome walking. Genomics 1995; 25:674-681.
77. Liu YG, Mitsukawa N, Oosumi T et al. Efficient isolation and mapping of Arabidopsis thaliana T-DNA insert junctions by thermal asymmetric interlaced PCR. Plant J 1995; 8:457-463.
78. González-Ballester D, de Montaigu A, Galván A et al. Restriction enzyme site directed amplification (RESDA)-PCR: A tool to identify regions flanking a marker DNA. Anal Biochem 2005; 340:330-335.
78. Harrison RW, Miller JC, D'Souza MJ et al. Easy gene walking. Biotechniques 1997; 22:650-653.
79. González-Ballester D. Genómica functional de la señalización por amonio y nitrato, y caracterización de genes para el transporte de amonio en Chlamydomonas. PhD Thesis 2005, (Universidad de Córdoba, Spain).
80. Peréz-Alegre M, Dubus A, Fernández E. REM1, a new type of long terminal repeat retrotransposon in Chlamydomonas reinhardtii. Mol Cell Biol 2005; 25:10628-10638.
81. Thyssen C, Hermes M, Sultemeyer D. Isolation and characterisation of Chlamydomonas reinhardtii mutants with an impaired CO_2-concentrating mechanism. Planta 2003; 217:102-112.
82. Polle JE, Kanakagiri SD, Melis A. Tla1, a DNA insertional transformant of the green alga Chlamydomonas reinhardtii with a truncated light-harvesting chlorophyll antenna size. Planta 2003; 217:49-59.

Chapter 8

Optimization of Recombinant Protein Expression in the Chloroplasts of Green Algae

Samuel P. Fletcher, Machiko Muto and Stephen P. Mayfield*

Abstract

Through advances in molecular and genetic techniques, protein expression in the chloroplasts of green algae has been optimized for high-level expression. Recombinant proteins expressed in algae have the potential to provide novel and safe treatment of disease and infection where current, high-cost drugs are the only option, or worse, where therapeutic drugs are not available due to their prohibitively high-cost to manufacture. Optimization of recombinant protein expression in *Chlamydomonas reinhardtii* chloroplasts has been accomplished by employing chloroplast codon bias and combinatorial examination of promoter and UTR combinations. In addition, as displayed by the expression of an anti-herpes antibody, the *C. reinhardtii* chloroplast is capable of correctly folding and assembling complex mammalian proteins. These data establish algal chloroplasts as a system for the production of complex human therapeutic proteins in soluble and active form, and at significantly reduced time and cost compared to existing production systems. Production of recombinant proteins in algal chloroplasts may enable further development of safe, efficacious and cost-effective protein therapeutics.

Introduction

Recombinant technologies have enabled rapid identification of proteins capable of providing an array of therapeutic functions. Complimented by advances in genomic and proteomic techniques, the number of therapeutic protein applications continues to grow as more advanced molecules and approaches are developed. Recombinant proteins (rP) offer great potential as therapeutic agents for a variety of human conditions, including those related to chronic disease, infectious agents and cancers. However, as protein identification and engineering techniques have advanced, the need for efficient and rapid production systems has emerged as a universal limitation in therapeutic protein production. A key consideration in the development of any new protein based drug is the inherent high cost of goods and capital investment associated with the production of these molecules. For example, monoclonal antibody production currently requires about $2 million per kg of product in capitalization costs,[1] while production costs often exceed $150/gm for antibodies prior to purification.[2] Moreover, many recombinant protein therapies require large amounts of protein per treatment, often in the range of grams per patient, which coupled with the high cost of goods makes some protein based therapies prohibitively expensive.

*Corresponding Author: Stephen P. Mayfield—Department of Cell Biology and The Skaggs Institute for Chemical Biology, The Scripps Research Institute, 10550 North Torrey Pines Road, La Jolla, California 92037, U.S.A. Email: mayfield@scripps.edu.

Transgenic Microalgae as Green Cell Factories, edited by Rosa León, Aurora Galván and Emilio Fernández.

Clearly a challenge exists to develop expression systems and appropriate methodologies to address the cost and demand issues, while still providing clinical grade protein therapeutics. The potential to generate novel and potentially superior medicines is now at hand, given the large annotated protein databases generated from precise screening techniques and genomic, molecular and cellular understanding of many diseases and pathological states. Herein, work to optimize rP expression in the chloroplast of green algae as a means to address the shortcomings in existing therapeutic protein availabilities and production systems is described. The ultimate goal is to develop algae for the expression of safe and effective rPs as important therapeutics to treat and cure debilitating human diseases.

Systems for the Expression of Recombinant Proteins

Currently, there are a number of protein expression systems available for the production of rPs, and each of these systems offers distinct advantages in terms of protein yield, ease of manipulation, and cost of operation.[2] Many therapeutic rPs today come from the culture of transgenic mammalian cells in fermentation facilities. Due to high capital and media costs, and the inherent complexity of these production systems, rPs produced in this manner are very expensive. Yeast and bacterial systems, while more economical in terms of media components, have several shortcomings in terms of rP expression, including an inability to efficiently produce properly folded full length molecules, as well as poor yields of more complex proteins.

In addition to these traditional systems, several groups have attempted to exploit the productivity of terrestrial plants for rP production. In such systems, the rP is synthesized within the plant cells and deposited into leaf or seed tissues. An array of rPs with potential therapeutic value have been produced in plants,[3-5] including functional antibodies,[6] vaccines,[7] enzymes,[8] hormones[9] and a variety of other proteins.[10] Protein production in photosynthetic cells is inherently less expensive than production attained by cell fermentation. Consequently, plants offer an attractive system for expression of recombinant proteins and perhaps the best economic alternative for the expression of complex multimeric proteins such as antibodies.[6]

While plants afford an economy of scale unprecedented in the pharmaceutical industry (one can plant thousands of acres in corn or tobacco, for example), there are several inherent drawbacks to this approach. First, the length of time required from the initial transformation event to having usable (mg to gram) quantities of antibody can take up to two years for crops such as tobacco, and over three years for species such as corn or banana. A second concern surrounding the expression of human therapeutics in crop plants, is the potential for gene flow (via pollen) to surrounding food crops,[11] as occurred between transgenic corn expressing *Bacillus thuringiensis* insecticidal proteins and native landraces.[12] Aside from concerns over gene-flow, many have expressed concerns about the food supply becoming contaminated with transgenic seeds expressing a human therapeutic. Thus, from both a time and regulatory perspective, food crops may not be ideal candidates for the expression of human therapeutic proteins.

Why Use Eukaryotic Algae for the Production of Recombinant Proteins?

As described above, plants are the most efficient system available to develop as a rP production platform. Plants utilize sunlight as their energy source and extract CO_2 from the air as their carbon source, and media and heating costs are significantly lower than bacterial, yeast and mammalian systems. As a system highly related to the plant family, algae are as efficient as terrestrial plants in the production of proteins, and theoretically can produce proteins at a fraction of the cost of traditional fermentation systems. Algae are also eukaryotes, meaning that unlike bacteria, they are efficient at producing complex proteins and have the machinery necessary to fold and assemble multi-component complexes into functional proteins. In addition, most plants and algae are generally regarded as safe (GRAS), posing little risk of viral, prion or bacterial endotoxin contamination in protein extracts.

Compared to land plants, algae like *Chlamydomonas reinhardtii* grow at a much faster rate, doubling cell number in approximately 8 hours under a 12 hour light, 12 hour dark regime. As *C. reinhardtii* propagates by vegetative division, the time from initial transformation to protein

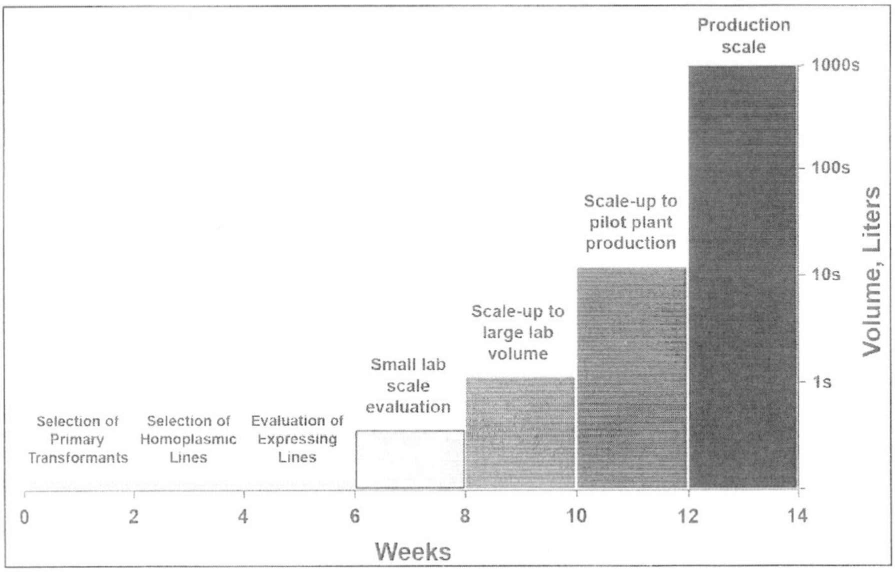

Figure 1. Theoretical scale-up from initial transformation of algae to large production volumes.

production is significantly reduced relative to plants, requiring as little as six weeks to evaluate protein production at flask scale, with the potential to scale up to 1000s of liters in a few months time (Fig. 1). *C. reinhardtii* also possesses a well characterized mating system, making it possible to generate algal strains expressing multi-subunit recombinant proteins (such as sIgA which contains four distinct peptides including heavy chain, light chain, J chain and secretory component, through matings between various transgenic algal lines, again in a very short period of time (3-4 weeks). The purification of recombinant proteins should also be greatly simplified relative to terrestrial plants as the cellular population of algae is uniform in size and type, hence, there is no gradient of recombinant protein distribution, simplifying the purification and reducing the amount of biomass which goes toward nonproductive ends. *C. reinhardtii* also has the ability to produce secreted proteins, potentially further reducing production costs.

Finally, algae like *C. reinhardtii* propagate by vegetative duplication and have no potential for gene transfer to food crops. *C. reinhardtii* is aquatic and can easily be grown in full containment in which the algae are sealed within closed plastic bags, again reducing any chance of environmental contamination. Growth in containment also assures that external contaminants, like pesticides or pollutants, do not get into the algal cultures to contaminate the protein drugs being produced; an important consideration for appropriate good manufacturing practices.

Expression of Recombinant Proteins in the *Chlamydomonas* Chloroplast

As listed in Table 1, a few attempts have been made to engineer chloroplasts of higher plants for the expression of therapeutic proteins and in some instances quite high levels of rP expression have been achieved in this organelle. There have been even fewer reports on the generation of transgenic algae for the expression of recombinant proteins even though green algae have served as a model organism for understanding everything from the mechanisms of light and nutrient regulated gene expression to the assembly and function of the photosynthetic apparatus.[14]

Table 1. Examples of recombinant protein expression in algae and plant chloroplasts

Host	Protein Products	Application	Refs.
Chlamydomonas reinhardtii	GFP, Lux (reporter gene)	Determining expression characteristics (codon usage, differing UTR sequences, etc.) in the chloroplast.	15,16
	HSV8 (antibody)	Pharmaceutical	17
Nicotiana tabaccum (tobacco)	Bacillus thuringiensis insecticidal protein	Agricultural application	18
	Glyphosate; PPT (herbicide) resistance	Agricultural application	19,20
	Somatotropin (human growth hormone)	Pharmaceutical	9
	Human serum albumin	Pharmaceutical	21
	Cholera toxin B subunit (vaccine)	Pharmaceutical	22
Lycopersicon esculentum (tomato)	Spectinomycin resistance gene, *aadA* (selectable marker)	Development of tomato plastid expression system	23
Solanum tuberosum L. cv. FL1607 (Potato)	GFP (reporter gene)	Development of potato chlorolplast expression system	24

Transformation of the *C. reinhardtii* chloroplast genome was first demonstrated in 1988,[25] and proceeds by site-specific recombination between homologous DNA sequences.[26] *C. reinhardtii* contains a single large chloroplast that occupies approximately 40-60% of the cell volume, and the chloroplast genome is a circular molecule of 200 kbp with each chloroplast containing 50-80 identical copies of the genome. Transformation events can be precisely targeted to any region in the chloroplast genome that contains a so-called "silent site" for transgene integration, and our lab has developed a number of such integration sites for recombinant gene integration. Selection of chloroplast transformants is provided either by cotransformation with DNA conferring resistance to antibiotics,[27,28] or through rescue of phototrophy.[25]

Expression of rPs in *C. reinhardtii* chloroplast has several advantages, including the rapid generation of stable transgenic lines, simple promoter and expression elements, relative rP stability with turnover rates similar to those of endogenous proteins and the potential for reproducibly high levels of recombinant protein accumulation.[15] In the case of human therapeutic proteins, the absence of glycosylation in the chloroplast may enhance the efficacy and safety of some therapeutic proteins. For example, extensive in-vitro and preclinical experiments demonstrate that aglycosylated recombinant antibodies retain biological activity, have similar clearance rates to glycosylated proteins, and do not ellicit unwanted immune responses.[29,30] The biological activity and clearance rates of rPs, glycosylated or aglycosylated, will be protein and case dependent, and these questions need to be addressed prior to identifying a system for transgenic production.

Expression of Reporter Genes in the Chlamydomonas *Chloroplast*

Reporter genes have been used to identify many of the characteristics of rP expression and accumulation in algal chloroplast, as these genes can provide the ability to monitor gene expression in a biological system as transient or stable visual markers. Conceptually, the use of reporter genes is ideal for monitoring protein accumulation and expression kinetics in the chloroplast. However, the use of traditional reporter genes such as β-glucuronidase (GUS) or chloramphenicol acetyltransferase in assaying protein expression from plastid promoters in tobacco and cultured cells has been problematic since protein expression quantitation requires the assay of reporter activity in cell extracts (as opposed to live cells). Low levels of protein accumulation also hindered the use of reporter genes in these plant expression systems.[31,32] Thus, a method to monitor plastid gene expression in real time, in vivo, would clearly be more convenient and reliable for observing rP expression and accumulation.

Until recently, the study of rP expression in the *C. reinhardtii* chloroplasts was limited, primarily due to poor expression of heterologous genes. Green fluorescent protein (GFP) from the jellyfish, *Aequorea victoria*, has been used in a variety of transgenic systems and has enabled researchers to monitor GFP-fusion proteins in numerous organisms. Using a codon optimized GFP reporter, high levels of recombinant protein accumulation were observed in the *C. reinhardtii* chloroplast, compared to a nonoptimized GFP counterpart.[15] Furthermore, GFP accumulation may be visualized in live cells, enabling in-situ observations of GFP-fusion rPs in intact cells. In addition, the recent development of a chloroplast codon optimized *luxAB* gene (*luxCt*) allowed for the visualization of chloroplast expressed rP accumulation, in situ, with relative ease.[16] Together, these codon optimized GFP and luxCt reporters have been instrumental in examining chloroplast gene expression, and continue to be valuable tools in the discovery and optimization of rP expression in algal chloroplasts.

Strategies for Increasing Recombinant Protein Expression in Algal Chloroplast

The Impact of Codon Optimization on Recombinant Protein Expression

Initial attempts to express mammalian recombinant proteins in *C. reinhardtii* chloroplasts were based on the expression of a single chain variable fragment (scFv) derived from a mouse monoclonal antibody raised against tetanus toxin. These initial attempts demonstrated that mammalian proteins could be expressed in algal chloroplasts, but the levels of protein expression achieved were very low, only a fraction of a percent of total protein. Examination of the codon usage in chloroplasts revealed that a number of codons were rarely used in photosynthetic genes, and this codon bias was not identified in the mammalian antibody initially expressed. To directly assay the impact codon bias might have on the expression of recombinant proteins in the *C. reinhardtii* chloroplast, GFP was synthesized with optimized codon usage to reflect the codon bias of major *C. reinhardtii* chloroplast encoded proteins. This chloroplast codon optimized GFP gene (*GFPct*) was tested against a GFP gene with the identical amino acid sequence but in noncodon bias (*GFPncb*). Both genes were cloned into expression cassettes under the control of the *rbcL* 5' and 3' UTRs. Homoplasmic strains were identified for each gene, and the accumulation of GFP was assayed. Using these two strains, it was apparent that GFP accumulation in chloroplasts transformed with the *GFPct* cassette accumulate approximately 80-100 fold more GFP than the *GFPncb* transformed strain, indicating that codon usage has a profound effect on the expression of heterologous proteins in the *C. reinhardtii* chloroplast.[15] Expression of the *GFPct* gene, under control of the *rbcL* 5' and 3' UTRs, produced GFP accumulation to about 1% of total soluble protein. These are levels sufficiently robust for the detection of fluctuations in gene expression due to environmental effects, making the *GFPct* gene a generally useful reporter gene for *C. reinhardtii* chloroplasts.

Based on the success of the *GFPct* gene in increasing protein accumulation, expression of monoclonal antibodies was reexamined by synthesizing a chloroplast codon optimized gene encoding a human IgA1 anti-herpes antibody. This large single chain protein contained the entire heavy chain coding region linked to the variable region of the light chain. Expression of this codon optimized, large single chain gene (*hsv8lsc*) driven by either the *atpA* or *rbcL* promoters and 5' UTRs resulted in the accumulation of a completely soluble protein product, again to about 1% of total soluble protein.[17] In addition, this large single chain antibody dimerized to form a correctly assembled, functional antibody capable of binding herpes simplex

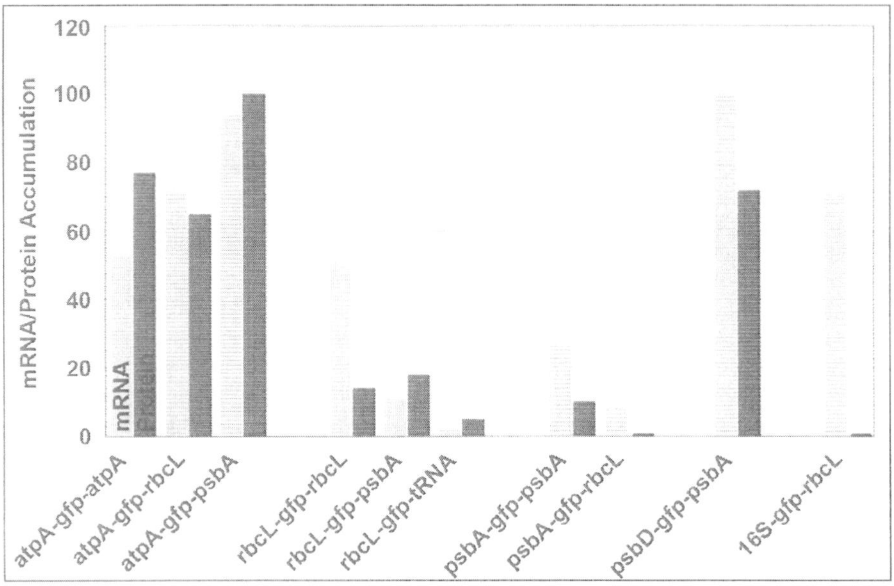

Figure 2. Analysis of GFP mRNA and protein accumulation. mRNA and protein levels were normalized to 16S and Rubisco, respectively. The data represent percentages of the highest levels of RNA and protein accumulation, psbD-gfp-psbA and atpA-gfp-psbA, respectively.

coat protein.[17] These data corroborated the idea that codon bias was essential to achieving good levels of protein accumulation in algal chloroplasts, and also established that algal chloroplasts have the capacity to synthesize complex human antibody molecules in a soluble and active form.

Impact of mRNA Accumulation on Recombinant Protein Expression

Accumulation of chloroplast proteins can be impacted at a variety of points during synthesis, including transcription, mRNA accumulation, translation or protein turn-over. Examining the expression of endogenous chloroplast genes under a variety of environmental conditions and developmental stages has shown that plastid protein accumulation is primarily determined by the rate of translation, as chloroplast mRNA accumulation is generally not rate limiting for plastid protein accumulation.[33] To identify if recombinant protein accumulation in *C. reinhardtii* was also independent of recombinant mRNA accumulation we examined mRNA and protein accumulation in a series of transgenic algae expressing *GFPct* driven by different chloroplast promoter and UTR combinations. The promoter and 5' UTRs of five *C. reinhardtii* chloroplast genes; *atpA*, *rbcL*, *psbA*, *psbD* and 16S *rRNA* were fused to the *gfpct* coding region followed by the 3' UTR of the *atpA*, *rbcL*, *psbA* or tRNAleu genes in a combinatorial fashion to generate an array of chimeric genes. Examination of mRNA accumulation from this set of reporter constructs showed that the different promoters and 5' UTRs each allowed for the accumulation of chimeric mRNAs, but at very different levels (Fig. 2). Accumulation of chimeric mRNAs was largely independent of the 3' UTR used, suggesting that mRNA accumulation is primarily determined by the promoter and 5' UTR.[34] Examination of the entire set of chimeric mRNAs, coupled with GFP analysis, showed a good correlation between mRNA and protein accumulation.[34] There were however, several notable exceptions to this correlation, and substantial variation between mRNA and protein accumulation for some promoter and UTR combinations (Fig. 2).

Impact of Translation on Recombinant Protein Expression

Although there was generally a fairly good correlation between chimeric mRNA accumulation and recombinant protein accumulation for the various GFP constructs, there were several examples where the correlation did not hold up. There was also a poor correlation between the amount of protein derived from chimeric mRNAs compared to the amount derived from endogenous mRNAs. Together these data show that translational efficiencies of chimeric mRNAs are much lower than translational efficiencies of endogenous chloroplast mRNAs.[34] This suggests that interactions of the coding region and UTRs can influence mRNA translation, and also infers that optimizing translation of chimeric mRNA could dramatically impact recombinant protein accumulation. There are a number of ways to impact translation of chimeric mRNAs including identification of optimal UTRs by a combinatorial approach, and also by introducing mutation that could potentially impact UTR/coding region interacting to increase translational efficiency.

Although translational efficiencies were very different between chimeric mRNA containing either the *psbA* or *psbD* 5' UTR, both chimeras retained light regulated translation in a similar manner to their endogenous counterparts, showing that the 5' UTR is sufficient to confer light regulated translation, and that translational efficiency and light regulated translation are not directly interconnected.[34]

Impact of Genetic Background on Recombinant Protein Accumulation

When the *luxct* gene was transformed into algal strains 137c or cc744, two very different luminescence outputs were detected. This result was unexpected, and it is unknown whether the difference in *luxct* expression between the two strains is due to differences in protein synthesis or protein turnover. Although it is not yet clear why the two strains show differences in accumulation of luciferase, it is clear that the luminescence output in each strain is proportional to the accumulation of luciferase protein, and thus, *luxct* is acting as a valid reporter of chloroplast gene expression in these two lines. The analysis of luciferase synthesis and turnover in both the 137c and cc744 strains is on going to determine the source of variation in gene expression between the two strains, but clearly these data show that the genetic makeup of a transgenic strain can have a profound impact on rP expression in chloroplasts.

Conclusion and Prospectus

Expression of Human Recombinant Proteins in C. reinhardtii *Chloroplast*

Through an analysis of the factors affecting GFP and Lux accumulation, high-levels of rP expression, and correct processing and assembly, have been achieved in *C. reinhardtii* chloroplasts. The optimization of rP expression in *C. reinhardtii* chloroplasts was accomplished, in part, by employing chloroplast codon bias for the rP genes, and by examining a number of promoter and UTR combinations to identify those that allowed for high levels of rP expression. Expression of rPs in *C. reinhardtii* chloroplasts was predominantly dependent upon the 5' UTR (compared to 3' UTR), and a fairly good correlation between mRNA accumulation and protein accumulation was observed. In addition, as displayed by the *hsv8lsc* antibody, *C. reinhardtii* chloroplasts are capable of correctly assembling functional protein complexes. These data establish algal chloroplasts as a system to synthesize complex human rP molecules in soluble and active form at significant levels.

Future Prospects for Therapeutic Protein Expression in Micro-Algae

Researchers have taken advantage of more than eighty years of genetics and thirty years of molecular biology to begin to engineer efficient expression of important therapeutic proteins in *C. reinhardti*. With opportunity for further development, rP expression in *C. reinhardtii* chloroplast has made rapid and substantial progress over a relatively short period of time compared to other recombinant protein expression systems. Algae are an attractive system for the expression of human therapeutic proteins given that they are eukaryotic and can thus assemble and fold complex mammalian proteins, they can be cultured in containment with relative ease allowing for good manufacturing practices, and they can be grown at a scale in facilities that allow for tremendous cost savings. Therapeutic protein expression in *C. reinhardtii* chloroplast will undoubtedly be optimized further as molecular techniques are advanced, as additional proteins are studied and preclinical and clinical trials are established to develop safe, efficacious and cost-effective protein therapeutic production.

Acknowledgements

This work was supported by Grants to SPM from The National Institute of Allergy and Infectious Diseases A1059614 and the Department of Energy DEFG0302ER15313.

References

1. UBS Investment Research: The Q Series™: The State of Biomanufacturing. 2003:1-52.
2. Dove A. Uncorking the biomanufacturing bottleneck. Nat Biotechnol 2002; 20(8):777-779.
3. Ma JK, Barros E, Bock R et al. Molecular farming for new drugs and vaccines. Current perspectives on the production of pharmaceuticals in transgenic plants. EMBO Rep 2005; 6(7):593-599.
4. Fischer R, Stoger E, Schillberg S et al. Plant-based production of biopharmaceuticals. Curr Opin Plant Biol 2004; 7(2):152-158.
5. Ko K, Koprowski H. Plant biopharming of monoclonal antibodies. Virus Res 2005; 111(1):93-100.
6. Hiatt A, Cafferkey R, Bowdish K. Production of antibodies in transgenic plants. Nature 1989; 342(6245):76-78.
7. Mason HS, Lam DM, Arntzen CJ. Expression of hepatitis B surface antigen in transgenic plants. Proc Natl Acad Sci USA 1992; 89(24):11745-11749.
8. Fletcher SP, Geyer BC, Mor TS. Translational control of recombinant human acetylcholinesterase accumulation in plants. submitted to the Journal of Plant Physiology
9. Staub JM, Garcia B, Graves J et al. High-yield production of a human therapeutic protein in tobacco chloroplasts. Nat Biotechnol 2000; 18(3):333-338.
10. Borisjuk NV, Borisjuk LG, Logendra S et al. Production of recombinant proteins in plant root exudates. Nat Biotechnol 1999; 17(5):466-469.
11. Ellstrand NC. When transgenes wander, should we worry? Plant Physiol 2001; 125(4):1543-1545.
12. Quist D, Chapela IH. Transgenic DNA introgressed into traditional maize landraces in Oaxaca, Mexico. Nature 2001; 414(6863):541-543.
13. Ma JK, Hikmat BY, Wycoff K et al. Characterization of a recombinant plant monoclonal secretory antibody and preventive immunotherapy in humans. Nat Med 1998; 4(5):601-606.
14. Harris E. The Chlamydomonas Sourcebook. In: Academic Press, Inc, 1989:780.
15. Franklin S, Ngo B, Efuet E et al. Development of a GFP reporter gene for Chlamydomonas reinhardtii chloroplast. Plant J 2002; 30(6):733-744.
16. Mayfield SP, Schultz J. Development of a luciferase reporter gene, luxCt, for Chlamydomonas reinhardtii chloroplast. Plant J 2004; 37(3):449-458.
17. Mayfield SP, Franklin SE, Lerner RA. Expression and assembly of a fully active antibody in algae. Proc Natl Acad Sci USA 2003; 100(2):438-442.
18. McBride KE, Svab Z, Schaaf DJ et al. Amplification of a chimeric Bacillus gene in chloroplasts leads to an extraordinary level of an insecticidal protein in tobacco. Biotechnology (NY) 1995; 13(4):362-365.
19. Ye GN, Hajdukiewicz PT, Broyles D et al. Plastid-expressed 5-enolpyruvylshikimate-3-phosphate synthase genes provide high level glyphosate tolerance in tobacco. Plant J 2001; 25(3):261-270.
20. Lutz KA, Knapp JE, Maliga P. Expression of bar in the plastid genome confers herbicide resistance. Plant Physiol 2001; 125(4):1585-1590.
21. Fernandez-San Millan A, Mingo-Castel A, Miller M et al. A chloroplast transgenic approach to hyper-express and purify Human Serum Albumin, a protein highly susceptible to proteolytic degradation. Plant Biotechnology Journal 2003; 1(2):71-79.
22. Daniell H, Lee SB, Panchal T et al. Expression of the native cholera toxin B subunit gene and assembly as functional oligomers in transgenic tobacco chloroplasts. J Mol Biol 2001; 311(5):1001-1009.
23. Ruf S, Hermann M, Berger IJ et al. Stable genetic transformation of tomato plastids and expression of a foreign protein in fruit. Nat Biotechnol 2001; 19(9):870-875.

24. Sidorov VA, Kasten D, Pang SZ et al. Technical Advance: Stable chloroplast transformation in potato: Use of green fluorescent protein as a plastid marker. Plant J 1999; 19(2):209-216.
25. Boynton JE, Gillham NW, Harris EH et al. Chloroplast transformation in Chlamydomonas with high velocity microprojectiles. Science 1988; 240(4858):1534-1538.
26. Newman SM, Boynton JE, Gillham NW et al. Transformation of chloroplast ribosomal RNA genes in Chlamydomonas: Molecular and genetic characterization of integration events. Genetics 1990; 126(4):875-888.
27. Fischer N, Stampacchia O, Redding K et al. Selectable marker recycling in the chloroplast. Mol Gen Genet 1996; 251(3):373-380.
28. Goldschmidt-Clermont M. Transgenic expression of aminoglycoside adenine transferase in the chloroplast: A selectable marker of site-directed transformation of chlamydomonas. Nucleic Acids Res 1991; 19(15):4083-4089.
29. Friend PJ, Hale G, Chatenoud L et al. Phase I study of an engineered aglycosylated humanized CD3 antibody in renal transplant rejection. Transplantation 1999; 68(11):1632-1637.
30. Simmons LC, Reilly D, Klimowski L et al. Expression of full-length immunoglobulins in Escherichia coli: Rapid and efficient production of aglycosylated antibodies. J Immunol Methods 2002; 263(1-2):133-147.
31. Daniell H, Vivekananda J, Nielsen BL et al. Transient foreign gene expression in chloroplasts of cultured tobacco cells after biolistic delivery of chloroplast vectors. Proc Natl Acad Sci USA 1990; 87(1):88-92.
32. Inada H, Seki M, Morikawa H et al. Existence of three regulatory regions each containing a highly conserved motif in the promoter of plastid-encoded RNA polymerase gene (rpoB). Plant J 1997; 11(4):883-890.
33. Eberhard S, Drapier D, Wollman FA. Searching limiting steps in the expression of chloroplast-encoded proteins: Relations between gene copy number, transcription, transcript abundance and translation rate in the chloroplast of Chlamydomonas reinhardtii. Plant J 2002; 31(2):149-160.
34. Barnes D, Franklin S, Schultz J et al. Contribution of 5'- and 3'-untranslated regions of plastid mRNAs to the expression of Chlamydomonas reinhardtii chloroplast genes. Mol Genet Genomics 2005:1-12.

CHAPTER 9

Phycoremediation of Heavy Metals Using Transgenic Microalgae

Sathish Rajamani, Surasak Siripornadulsil, Vanessa Falcao, Moacir Torres, Pio Colepicolo and Richard Sayre*

Abstract

Microalgae account for most of the biologically sequestered trace metals in aquatic environments. Their ability to adsorb and metabolize trace metals is associated with their large surface:volume ratios, the presence of high-affinity, metal-binding groups on their cell surfaces, and efficient metal uptake and storage systems. Microalgae may bind up to 10% of their biomass as metals. In addition to essential trace metals required for metabolism, microalgae can efficiently sequester toxic heavy metals. Toxic heavy metals often compete with essential trace metals for binding to and uptake into cells. Recently, transgenic approaches have been developed to further enhance the heavy metal specificity and binding capacity of microalgae with the objective of using these microalgae for the treatment of heavy metal contaminated wastewaters and sediments. These transgenic strategies have included the over expression of enzymes whose metabolic products ameliorate the effects of heavy metal-induced stress, and the expression of high-affinity, heavy metal binding proteins on the surface and in the cytoplasm of transgenic cells. The most effective strategies have substantially reduced the toxicity of heavy metals allowing transgenic cells to grow at wild-type rates in the presence of lethal concentrations of heavy metals. In addition, the metal binding capacity of transgenic algae has been increased five-fold relative to wild-type cells. Recently, fluorescent heavy metal biosensors have been developed for expression in transgenic Chlamydomonas. These fluorescent biosensor strains can be used for the detection and quantification of bioavailable heavy metals in aquatic environments. The use of transgenic microalgae to monitor and remediate heavy metals in aquatic environments is not without risk, however. Strategies to prevent the release of live microalgae having enhanced metal binding properties are described.

Metals in the Environment

Living organisms require a number of essential metals (Cu, Mn, Co, Zn, Ni, Fe, Mo, Ca, Na, K, P) for general metabolism. These essential metals function as cofactors in enzymes, mediate redox reactions (photosynthetic and respiratory electron transfer chains), and participate in protein-protein structural interactions (e.g., zinc finger proteins). Many of these metals are present at very low abundance (<0.1%) in the earth's crust or are biologically unavailable due to their low solubility. As a result, energy often must be expended to import metals into cells or to convert them to more soluble forms.[1,2]

*Corresponding Author: Richard Sayre—Department of Plant Cellular and Molecular Biology, Ohio State University, Columbus, Ohio, 43210, U.S.A. Email: sayre.2@osu.edu

Transgenic Microalgae as Green Cell Factories, edited by Rosa León, Aurora Galván and Emilio Fernández.

Significantly, many nonessential heavy metals compete with essential metals for uptake into cells.[1-4] Several of these nonessential metals are toxic at very low concentrations (1.0 ppb to 1.0 ppm). Furthermore, at high concentrations even essential metals may be toxic. The molecular mechanisms by which metals damage cells include formation of nonfunctional protein-metal adducts, alterations in the redox state of cells, the generation of toxic free radicals and reactive oxygen species, and direct damage to DNA.[5,6] Acute or chronic exposure to toxic levels of metals can lead to the development of a variety of pathologies and cause reductions in agricultural yields.[5,6]

Human activities have had substantial impacts on the levels of bioavailable heavy metals in the environment. Nearly twenty years ago, it was estimated that global releases of lead, mercury, cadmium, copper and zinc from activities such as mining, smelting, coal combustion, plating and refuse disposal were 1,160; 11; 30; 2,150; and 2,340 thousand metric tons/year, respectively.[6] Since metals are elements they can not be chemically degraded. As a result, they accumulate in food chains and in the environment. The only effective strategies to remediate metal contaminated sites are to sequester the metals in biologically unavailable complexes or enclosures or to remove them from the contaminated site.

In terrestrial environments, bacteria, fungi and the primary producers (plants and algae) play the predominant role in the biogeochemical cycling of heavy metals. In aquatic environments algae play a key role in the biogeochemical cycling of metals and their accumulation in sediments. Metals sequestered by microalgae are a major contributor to the metal load of the water column as well as to the metal content of sediments.[7-9] In marine environments the sedimentation of phytoplankton during algal blooms has been shown to result in substantial (20-75%) reductions in the metal load of the water column and metal deposition in sediments.[10] This impact of phytoplankton on the biogeochemical cycling of metals in aquatic environments can be attributed both to the high abundance of microalgae at the base of the food chain and to the unusually high metal binding capacities of many microalgal species which equal or exceed the metal binding capacity of many commercial ion exchange resins.[11] The smallest microalgae, having the largest surface to volume ratios, are often the most effective at sequestering metals.[12]

Given these attributes, it is not unexpected that microalgae have received considerable attention for the treatment and remediation of metal contaminated wastes and sediments.[11,13] In the following sections, we review the biology of metal binding, uptake and sequestration by microalgae. We also describe recent progress in engineering microalgae to enhance their metal binding properties for the remediation of metal contaminated wastes and sediments as well as for monitoring bioavailable levels of metals in the environment.

The Role of the Algal Cell Wall in Heavy Metal Binding and Tolerance

At the interface between the physical environment and the cytoplasm is the cell wall. The cell wall is the first line of defense against toxic heavy metal poisoning. Microalgal cell walls typically have very high heavy metal binding capacities (0.10 g metal:g dry weight algae) effectively buffering the plasmamembrane from potentially toxic levels of metals.[8,9,11,14-18] The cell walls of *Chlamydomonas reinhardtii* have at least three different classes of functional heavy metal binding groups. These functional groups can be chemically distinguished by their pK_as (~3.5, ~4.5 and ~9.5).[14] The acidic and basic metal binding groups have high functional group acidities of ~ 3.5 mol $kGDW^{-1}$, and ~ 2 mol $kGDW^{-1}$, respectively, for a total functional group acidity of 5.5 mol $kGDW^{-1}$. Thus, the potential cadmium binding capacity of the cells is 61.6 gm/kGDW (6%). However, empirical studies under controlled pH and ionic strength indicate that the amount of sorbed cadmium may be as little as 1% of the available functional group sites.[14] These results suggest that their metal binding sites may be heterogeneous. Further evidence in support of the heterogeneity of metal binding sites has come from metal titration studies of Chlamydomonas cells at various pHs. These studies indicated that the binding of some (Cd) but not all (Au) metals is reversible as a function of pH.[8,13,14] For example, cadmium is bound most effectively at pH 8.0 and is quantitatively released from cell walls at low pH (2.0), while gold binding is pH independent and is not released at low pH (≥2.0). Spectroscopic analysis of the metal binding groups indicates that a variety of functional groups participate in metal binding. FTIR studies indicated that carboxylate-groups are involved in the pH-dependent binding of cadmium, however, X-ray fine structure spectroscopy (EXAFS) studies indicated that both carboxylate and sulfate groups bind cadmium.[13,14,19] This was not unexpected since Chlamydomonas cell wall glycoproteins contain 1% to 4% (g/g) sulfate as sugar-sulfate esters.[18] It has been proposed that these sulfated oligosaccharides protect the cell against desiccation, however, they also may play a substantive role in metal binding. Interestingly, comparative metal toxicity studies using walled and cell wall-less mutants (CC-425) of Chlamydomonas demonstrate that the I_{50} for growth in the presence of cadmium is 3-4 fold greater for walled cells than for wall-less cells.[17] These results clearly demonstrate an important role for the cell wall in buffering the toxic effects of heavy metals. As described below, the expression of heavy metal binding proteins at the interface between the cell wall and the plasmamembrane can substantially increase the metal binding capacity of Chlamydomonas. These results suggest that engineering cell wall proteins to increase their metal binding capacity could be an effective strategy to enhance metal binding specificity and capacity.

The Plasma Membrane and Heavy Metal Flux

One of the more active areas of research in heavy metal metabolism has been the identification and characterization of plasmamembrane metal transporters. Until recently, little was known about the biochemistry or molecular biology of metal import into Chlamydomonas. This is surprising since in many respects Chlamydomonas is an ideal single cell system to study heavy metal transport. Recently, the molecular mechanisms for metal ion transport have been investigated in Chlamydomonas by a variety of experimental approaches including functional genomic studies, biochemical and bioinformatics approaches. With the completion of the sequencing of the Chlamydomonas genome it has been possible to mine the database for conserved metal ion transporter genes present in other organisms. At least eleven unique gene families known to encode metal ion transporters are represented in the Chlamydomonas genome.[20] Interestingly, recent analyses have indicated that the full complement of metal ion transporter families present in Chlamydomonas is most similar to that of yeast (*Saccharomyces cerviseae*) although many transporter families are shared with plants, animals and bacteria.[20]

To date, there have been few functional studies on metal transporters in Chlamydomonas. Recently, however, it was demonstrated that a Chlamydomonas analogue of the Nramp family of metal ion transporters, DMT1, mediates Mn, Fe, Cd and Cu uptake but not Zn transport.[21] In addition, a novel green algal, periplasmic metal transporter known as Fea1 or the H43 protein has been shown to transport Fe but not other metals (Cd, Cu, Co or Mn) in yeast complementation experiments.[22] The Fea1 protein is expressed only under stress conditions including, exposure to toxic levels of cadmium, iron deficiency and high CO_2 concentrations (3%).

One successful approach to enhance the metal binding capacity of Chlamydomonas has involved the expression of synthetic genes encoding plasma membrane anchored metal-binding proteins exposed to the periplasmic space. Such a protein construct is shown in Figure 1. Various metallothionein-II (MT) polymers (1-5 MT domains), fused to the C-terminus of a low CO_2-induced plasmamembrane protein, were expressed in transgenic Chlamydomonas.[23,24] Metallothionein has a high specificity and binding affinity for Zn, Cd, Pb, Cu, Hg, Ag and Au.[25,26] Seven to eight atoms of metal are bound per MT holoprotein. Significantly, transgenic cells expressing these metal binding fusion proteins grew at normal (wild-type) rates in the presence of lethal concentrations of cadmium (120 μM for wall-less cells) and accumulated five-fold more Cd than wild-type cells.[24] These results suggested that the periplasmic MT domain buffered extra-cellular cadmium thus reducing its toxicity. Overall, these results

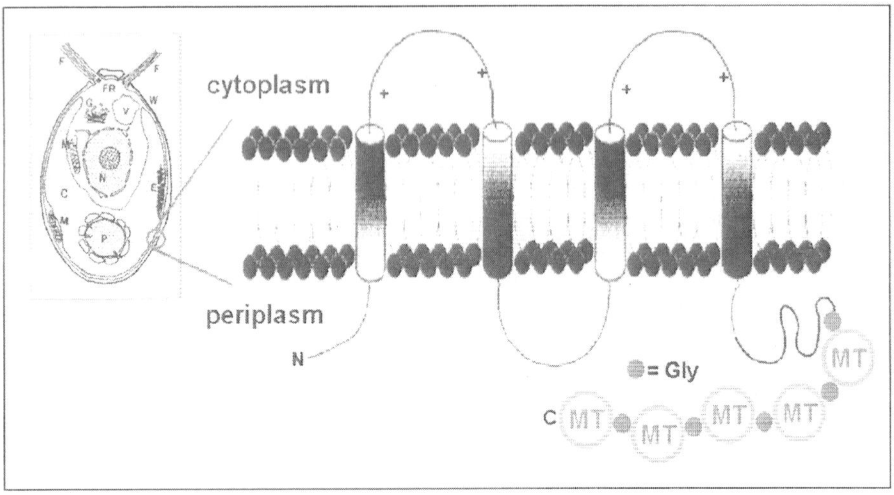

Figure 1. Model of the plasmamembrane-metallothionein (MT) fusion protein for enhanced heavy metal binding in transgenic Chlamydomonas.[23,24]

demonstrated that the heavy metal tolerance and accumulation can be altered through selective expression of transgenes in the periplasmic space of algae. Currently, cells expressing the periplasmic localized metallothionein are being tested for recovery of heavy metals released from contaminated sediments by in situ sonnication.[27] Preliminary results indicate that live transgenic Chlamydomonas cells expressing the periplasmic metallothionein constructs have a two-fold greater capacity to sequester heavy metals released from sonnicated sediments than do wild-type cells.

Heavy Metal Metabolism in the Cytoplasm of Algae

One the best characterized responses to toxic heavy metal exposure in microalgae is the induction of phytochelatin synthesis. Many microalgal species respond to heavy metal stress by synthesis of phytochelatins.[28-31] Phytochelatins are glutathione polymers, 2-9 units in length, which have high selectivity for only a few heavy metals including copper, cadmium, lead, mercury, zinc, silver and gold.[26] The genes encoding phytochelatin synthase have been isolated from yeast and Arabidopsis.[32-34] Biochemical studies have demonstrated that the synthesis of phytochelatins is regulated by the availability of heavy metal substrates, i.e., glutathione-heavy metal adducts activate synthesis of phytochelatins.[34] Once synthesized, phytochelatins are sequestered in vacuoules but also may be exported from some algae.[29,31,33,35] Regardless, intracellular heavy metal concentrations increase several fold following induction of phytochelatin synthesis.[17,29,31]

While algae clearly synthesize phytochelatins to sequester toxic heavy metals it also is evident that phytochelatins do not provide complete protection against high concentrations of toxic heavy metals. Recent studies using transgenic algae that constitutively expressed a foreign class II metallothionein demonstrated that expression of recombinant metallothioneins in the cytoplasm provided only limited protection against cadmium toxicity.[15-17] Chlamydomonas cells expressing a chicken metallothionein-II in the cytoplasm had two-fold higher growth rates than wild-type cells when grown in the presence of 50 μM cadmium but did not accumulate substantially more Cd than wild-type cells.

The heavy metal tolerance and binding capacity of Chlamydomonas has also been successfully enhanced through expression of foreign genes encoding enzymes mediating the synthesis of small molecular weight molecules that provide protection against heavy metal mediated damage. Siripornadulsil et al (2002) introduced genes encoding (1) pyrroline 5 carboxylate synthase (P5CS), which catalyzes the first-dedicated and rate-limiting step in proline synthesis, and (2) the *HAL2* gene product, which regulates cysteine synthesis into the nuclear genome of Chlamydomonas.[19,36-38] In some plant species, free proline levels are elevated in response to heavy metal stress.[38] Elevated proline is associated with increased heavy metal tolerance. In contrast to plants, free proline levels in Chlamydomonas are not affected by exposure to Cd concentrations sufficient to induce phytochelatin synthesis.[19,29,31] In transgenic cells expressing the *P5CS* gene free proline levels were elevated two fold relative to wild-type cells. Significantly, expression of the *P5CS* gene increased the metal binding capacity of transgenic algae by four-fold relative to wild-type cells when cells were grown in toxic concentrations (≥50 μM) of cadmium.[23] This increase in heavy metal binding capacity was associated with a similar four-fold increase in the ratio of reduced to oxidized glutathione. As shown by EXAFS studies, increased levels of reduced glutathione were correlated with increased Cd-phytochelatin levels. The two fold increase in proline levels in transgenic cells were also correlated with a two fold reduction in the products of ROS-mediated lipid peroxidation. These results were consistent with a proline-mediated quenching of reactive oxygen species induced by Cd stress.[19]

Similar to *P5CS* expression, expression of the *HAL2* gene in transgenic microalgae increased the heavy metal carrying capacity of the algae. The Cd capacity of *HAL2* transgenic cells was 2.5-fold greater than wild-type cells when grown in the presence of cadmium concentrations known to induce phytochelatin synthesis.[39] Since the *HAL2* gene regulates cysteine synthesis, expression of this enzyme presumably results in a more reducing (reduced glutathione) environment in the cell. EXAFS studies confirmed that the additional Cd present in transgenic cells was complexed to either glutathione or phytochelatins.[19]

Algal Heavy Metal Biosensors

To assess the potential biological impact of heavy metals in the environment it is critical to know both the total and the biologically available levels of heavy metals. The total heavy metal content is most frequently determined by atomic absorption or mass spectroscopy of acid solubilized samples and substrates. Bioavailable levels of heavy metals can indirectly be determined by similar analyses of the total metals present in an organism. These bioassays are destructive, however, and may be biased by the differential ability of different organisms to bioaccumulate metals. An alternative and nondestructive approach to monitor bioavailable heavy metals is to use bioindicator species.[40-44] These species are typically hypersensitive to heavy metals and have readily assayable traits (e.g., growth rate) that respond to heavy metals in a quantifiable manner.

Recently, Rajamani et al (2006) developed a Fluorescence Resonance Energy Transfer (FRET) based heavy metal biosensor that has been expressed in Chlamydomonas.[45] FRET is a distance dependent phenomenon involving a donor and an acceptor fluorescent molecule, which are typically modified versions of the jellyfish green fluorescent protein (GFP). FRET occurs when the distance between a donor-acceptor pair is <10 nm. Under suitable conditions, the excited donor molecule transfers its excitation energy to a lower energy acceptor molecule, resulting in decreased donor fluorescence and increased acceptor fluorescence.[46,47] FRET has been shown to be useful for monitoring distance changes in biological macromolecular systems and has found numerous applications in structural biology and biochemistry.[47,48] With the advent of GFP and its variants, FRET based biomonitoring systems have found use in both in vitro and in vivo studies.[49-51] Miyawaki et al, (1997) first demonstrated that GFP and its variants could be used for developing FRET based biosensors. They developed a calcium biosensor by tandem fusion of cyan fluorescent protein (CFP)-calmodulin-calmodulin binding protein and M13-yellow fluorescent protein (YFP) for monitoring calcium signaling.[52] Since then GFP variants, particularly CFP-YFP FRET pairs have been shown useful in several intramolecular and intermolecular FRET studies.[53]

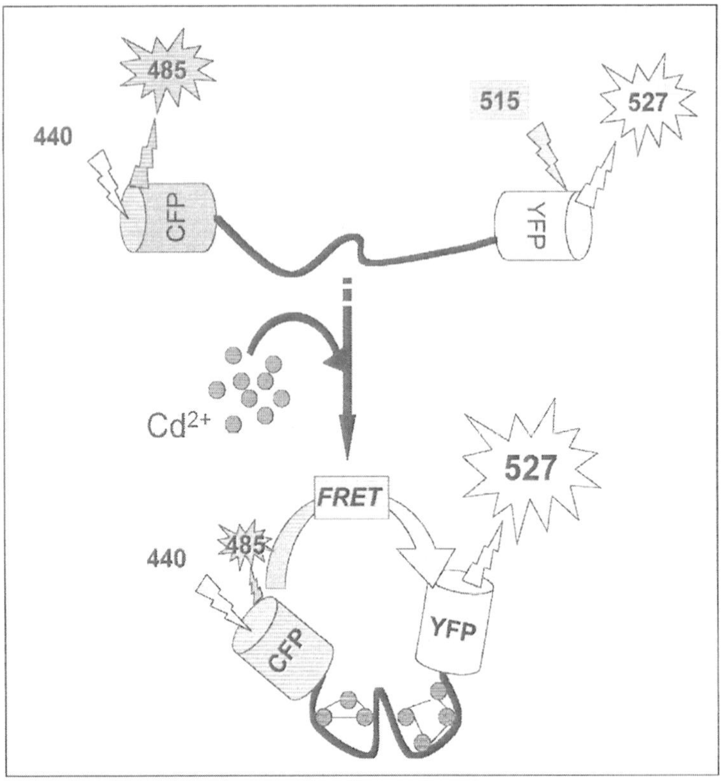

Figure 2. Schematic representation of Cd^{2+} binding to FRET biosensor CMY. CMY was made by tandem fusion of CFP (λ_{ex}- 440 nm, λ_{em}- 485 nm) with N-terminus of MT and C-terminus of MT to YFP (λ_{ex}- 515 nm,λ_{em}- 527 nm). β-domain (N-terminus) of MT binds 3 equivalents of divalent metal ion and α-domain (C-terminus) binds 4 equivalents of divalent metals. In the unbound (free) state MT is relaxed and the fluorophores attached to MT are farther from each other (≥100 Å). Bound MT has a closed conformation bringing the flurophores (CFP and YFP) closer (<70 Å) together resulting in increased FRET efficiency.

The FRET-based heavy metal biosensor expressed in Chlamydomonas was constructed by fusing a metallothionein II gene between genes encoding CFP and YFP. It had previously been demonstrated that metallothionein can be fused to the bulky fluorophores CFP and YFP (28kD each) to study metal release during nitric oxide signaling.[54] Chlamydomonas codon-optimized and functionally improved variants of CFP and YFP were utilized for creating a FRET-based heavy metal biosensor. In addition, mutations were engineered into CFP and YFP to facilitate faster protein maturation and to reduce protein-protein dimerization.[55,56] The resulting FRET construct was designated CMY. The scheme shown in Figure 2 depicts the functioning of the CMY, heavy metal biosensor.

For in vitro studies the CMY fusion protein was purified from bacteria as a maltose binding protein - CMY fusion protein. Various metals that bind to MT-II (Cd^{2+}, Pb^{2+}, Zn^{2+}) and other nonbinding metals (Na^{+}, Mg^{2+}) were tested for their ability to induce FRET. The CMY construct was shown specifically to respond to Cd^{2+}, Pb^{2+}, and Zn^{2+}, but not Na^{+} and Mg^{2+} by an increase in the YFP/CFP fluorescence ratio (FRET). Figure 3 shows the concentration-dependent increase in FRET ratio (527 nm/485 nm) with Cd^{2+} and Pb^{2+}, while Mg^{2+} had no effect on the FRET ratio. As little as 1.0 μM cadmium could be detected as a shift in the YFP/CFP ratio under the conditions tested.

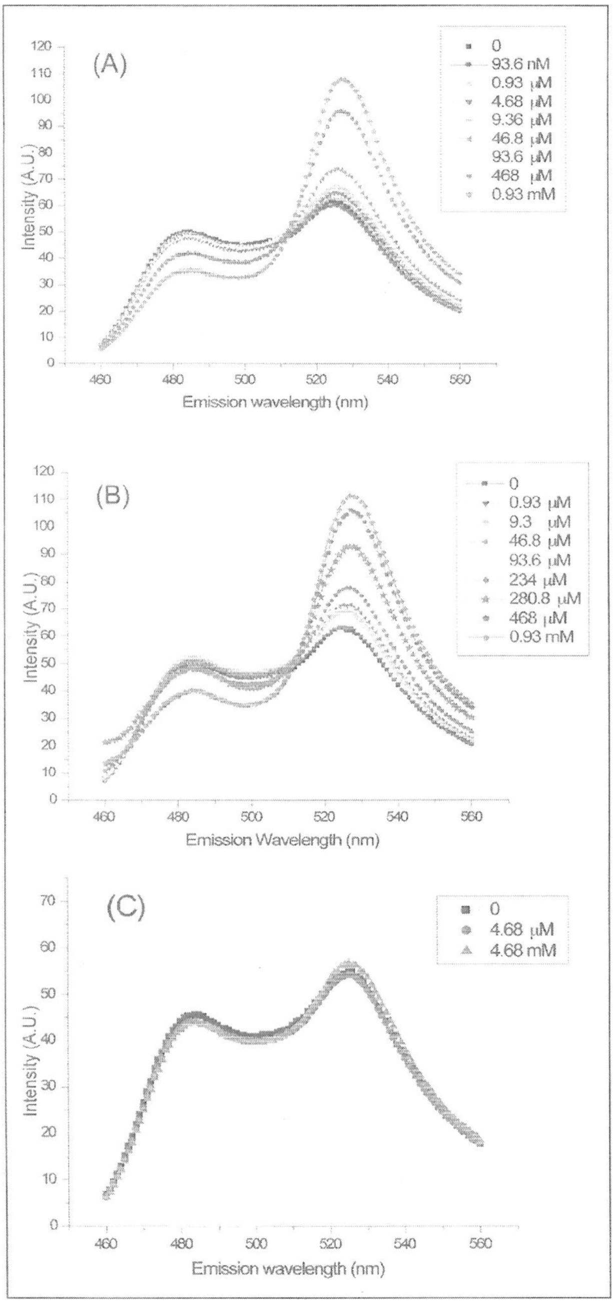

Figure 3. Concentration dependent increase in FRET ratio (527 nm/485 nm) seen with increased concentrations of Cd^{2+} (A) and Pb^{2+} (B), but no apparent FRET ratio change with excess (4.68 mM) Mg^{2+} (C). FRET measurements were carried out at room temperature, immediately after mixing protein with metal solutions. Measurement parameters: λ_{ex} - 440 nm, excitation slit - 5 nm, emission slit - 5 nm. Metal binding studies were carried out at a CMY protein concentration of 325 nm in 50 mM HEPES pH 7.4.

FRET studies with CFP-YFP pairs generally utilize excitation wavelengths between 430-440 nm. To standardize transgenic algal CMY (in vivo) fluorescence measurements and to reduce the absorbance of light used to excite CFP (at 440 nm) by chlorophyll, an excitation wavelength of 386 nm was experimentally determined to be most effective. We observed that algal cells exhibited fast kinetics (sub-second) for heavy metal uptake as indicated by the rate of increase in the YFP/CFP fluorescence emission ratio at threshold concentrations of heavy metals. Transgenic cells expressing the CMY heavy metal biosensor can also be readily imaged in a background of wild-type cells for easy monitoring (Fig. 4A). As expected, metals (Na and Mg) that do not bind to metallothionein did not induce a shift in the YFP/CFP fluorescence emission ratio, whereas metals that bind to metallothionein (Pb, Cd, Hg, Cu) induced a shift in the YFP/CFP fluorescence emission ratio (Fig. 4B). Interestingly, the threshold heavy metal concentration that induces a shift in the YFP/CFP ratio in CMY expressing cells was approximately 100-fold less than that observed for the isolated protein. It is hypothesized that chemical potential of free heavy metals in the cytoplasm is very low. This may be due to limitations in metal transport into the cell as well as to sequestration of heavy metals in the cytoplasm potentially as glutathione conjugates.

Application of Engineered Algae for Bioremediation: The Risks and Benefits

The application of single-celled microalgae for heavy metal bioremediation has risks as well as benefits that will require clever engineering and biological considerations to preclude the escape of transgenic organisms. The release of live transgenic microalgae with enhanced heavy metal binding capacities into the environment has the potential to cause greater harm than good. Algae with enhanced heavy metal capacities could accelerate the biogeochemical cycling of heavy metals and their accumulation in food chains. Therefore, multiple independent physical and biological barriers would need to be stacked to preclude the release of live transgenic algae having enhanced heavy metal binding capacities in any bioremediation application. Physical containment of transgenic microalgae would provide an effective barrier to prevent their escape but would also limit direct access of the algae to contaminated materials potentially reducing their effectiveness to compete with other metal-binding factors present in sediments.

Heavy metals could be recovered from contaminated sediments by in situ solubilization of metals using acetic acid. Using a grid of injection pumps the heavy metals present in contaminated sediments would be solubilized using weak acetic acid. The heavy metal contaminated solutions would then be recovered using a network of recovery wells surrounding the injection sites. The metal contaminated acetic solutions could then be used to augment the heterotrophic growth of Chlamydomonas cultures as well as deliver solubilized metals. Chlamydomonas cells engineered for enhanced metal binding capacity would be contained in closed chambers and the solutions containing solubilized metals would be exchanged across semi-permeable membranes that would preclude the release of transgenic cells. To insure that any potential algal escapees from the containment system were nonviable, multiple mutations would be introduced into the host strain that would preclude the growth of the transgenic algae or exchange of genetic material with nontransgenic algae in the wild. Examples of such mutations include the *nit1-30* and *pf-14* mutations that preclude growth on nitrate and prevent mating due to paralyzed flagella, respectively.[57] Deletion of the chloroplast-encoded and maternally-inherited, *psbA* gene which is required for photosynthesis would preclude autotrophic growth. Since the *psbA* gene is maternally inherited, a *psbA* lesion in a plus (maternal) mating type would insure transfer of the lesion to any offspring of a possible mating.[57] Mutants containing *psbA* lesions can be cultured heterotrophically, however, on acetate. Triple mutant strains containing a *psbA* deletion and the *nit1-30* and *pf-14* alleles would therefore require acetate and ammonia to survive and would be unable to mate. Finally, post treatment sterilization of the waste water effluent would further eliminate possible escape of transgenics.

An alternative strategy would be to use nonviable transgenic algae having enhanced metal binding capacity and selectivity. This strategy may hold the greatest promise for the development

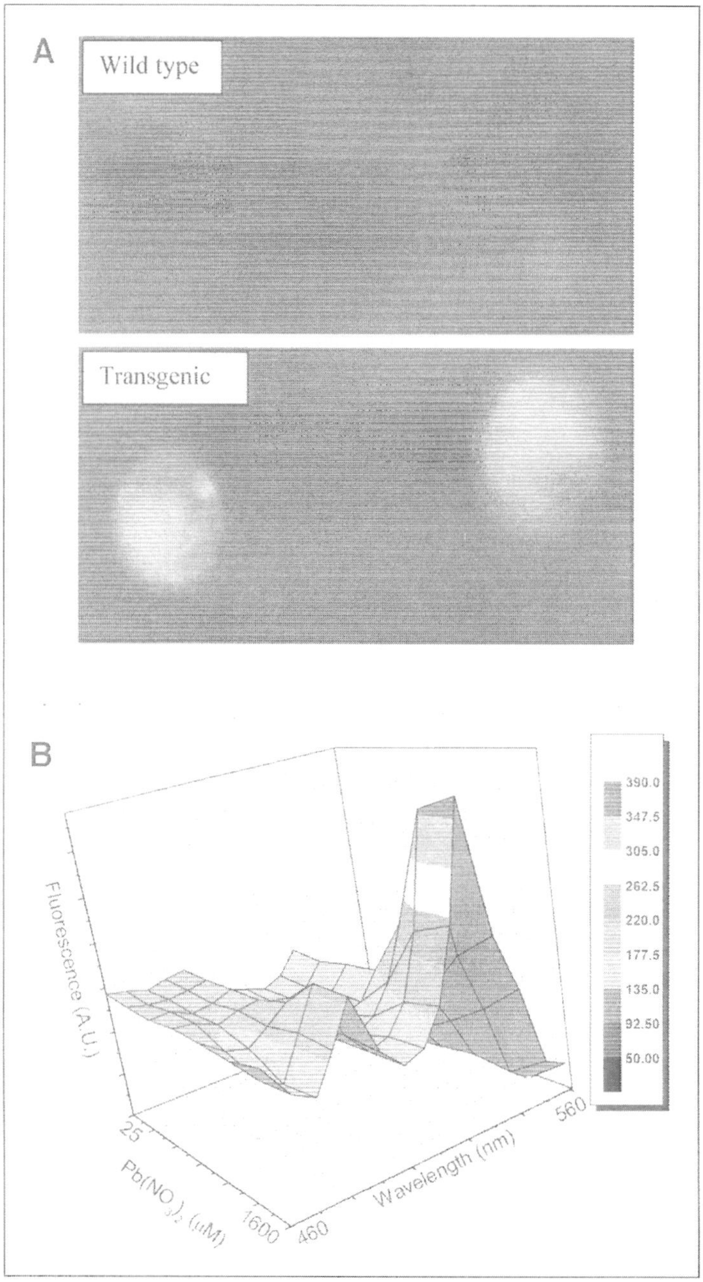

Figure 4. A) Detection of CMY fluorescence in transgenic algae by fluorescence imaging microscopy (40 x magnification). Transgenic Chlamydomonas expressing CMY showed higher fluorescence relative to the wild-type cells. B) Lead (Pb) dependent increase in the YFP (527 nm)/CFP (485 nm) fluorescence ratio in transgenic algae expressing the CMY biosensor. Algal cells at a cell density of (4×10^5 cells/ml) were exposed to lead nitrate at the designated concentrations. The color scale is relative fluorescence intensity.

of enhanced metal recovery and bioremediation of aqueous systems. Currently, improvements in the noninvasive and nondestructive quantification of bioavailable levels of metals are under development and hold promise for rapid and efficient monitoring of metals in aquatic environments. Given the central role microalgae play in the cycling of metals in aquatic environments and their exceptional abilities to bind and sequester metals it is evident that both wild and transgenic microalgae have promise as tools to recover and monitor metals.

References

1. Guerinot ML. The ZIP family of metal transporters. Biochim Biophys Acta 2000; 1465:190-198.
2. Williams LE, Pittman JK, Hall JL. Emerging mechanisms for heavy metal transport in plants. Biochim Biophys Acta 2000; 1465:104-126.
3. Cohen CK, Fox TC, Garvin DF et al. The role of iron-deficiency stress responses in stimulating heavy metal transport in plants. Plant Physiol 1998; 116:1063-1072.
4. Pence NS, Larsen PB, Ebbs SD et al. The molecular physiology of heavy metal transport in the Zn/Cd hyperaccumulator Thlaspi caerulescens. Proc Nat Acad Ssi 2000; 97:4956-4960.
5. Logan TJ, Traina SJ. Trace metals in agriculture. In: Allen HE, Perdue EM, Brown DS, eds. Metals in Groundwater. Chelsea, MI: Lewis Pub, 1993:309-349.
6. Nriagu JO, Pacyna JM. Quantitative assessment of worldwide contamination of the air, water and soils by trace metals. Nature 1988; 333:134-139.
7. Harris PO, Ramelow GJ. Binding of metal ions by particulate biomass derived from Chlorella vulgaris and Scenedesmus quadracauda. Environ Sci Technol 1990; 24:220-228.
8. Sandau E, Sandau P, Pulz O et al. Heavy metal sorption by marine algae and algal by-products. Acta Biotechnol 1996; 16:103-119.
9. Xue HB, Stumm W, Sigg L. The binding of heavy metals to algal surfaces. Wat Res 1988; 22:917-926.
10. Luoma SN, van Geen A, Lee BG et al. Metal uptake by phytoplankton duringa bloom in South San Francisco Bay: Implications for metal cycling in estuaries. Limnol. Oceanogr 1998; 43:1007-1016.
11. Mehta SK, Gaur JP. Use of microalgae for removing heavy metal ions from wastewater: Progress and prospects. Crit Rev Biotechnol 2005; 25:113-152.
12. Khoshmanesh A, Lawson F, Prince IG. Cell surface area as a major parameter in the uptake of cadmium by unicellular green microalgae. Chem Engineer J 1997; 65:13-19.
13. Tuzun I, Bayramoglu G, Yalcin E et al. Equilibrium and kinetic studies on Biosorption of Hg(II), Cd(II) and Pb(II) ions onto microalgae, Chlamydomonas reinhardtii. J Environ Manag 2005; 77:85-92.
14. Adhiya J, Cai XH, Sayre RT et al. Binding of aqueous cadmium by the lyophilized biomass of Chlamydomonas reinhardtii. Colloids and Surfaces A: Physicochemical and Engineering Aspects 2002; 210:1-11.
15. Cai XH, Logan T, Gustafson T et al. Applications of eukaryotic algae for the removal of heavy metals from water. Molecular Marine Biology and Biotechnology 1995; 4:338-344.
16. Cai XH, Brown C, Adihya J et al. Heavy metal binding properties of wild type and transgenic algae (Chlamydomonas sp.). In: Le Gal Y, Halvorson H, eds. New Developments in Marine Biotechnology. Plenum Press, 1998:199-202.
17. Cai XH, Brown C, Adihya C et al. Growth and heavy metal binding properties of transgenic algae (Chlamydomonas reinhardtii) expressing a foreign metallothionein gene. Int J Phytoremediation 1999; 1:53-65.
18. Roberts K, Gay MR, Hills GJ. Cell wall glycoproteins from Chlamydomonas reinhardtii are sulphated. Physiol Plant 1980; 49:421-424.
19. Siripornadulsil S, Traina S, Verma DP et al. Molecular mechanisms of proline-mediated tolerance to toxic heavy metals in transgenic microalgae. Plant Cell 2002; 14:2837-2847.
20. Hanikenne M, Kramer U, Demoulin V et al. A comparative inventory of metal transporters in the green alga Chlamydomonas reinhardtii and the red alga Cyanidioschizon merolae. Pl Physiol 2005; 137:428-446.
21. Rosakis A, Koster W. Divalent metal transport in the green microalga, Chlamydomonas reinhardtii is mediated by a protein similar to prokaryotic Nramp homolgues. BioMetals 2005; 18:107-120.
22. Rubinelli P, Siripornadulsil S, Gao-Rubinelli F et al. Cadmium and iron-stress inducible gene expression in the green alga Chlamydomonas reinhardtii: Evidence for H43 protein function in iron assimilation. Planta 2002; 215:1-13.
23. Siripornadulsil S, Traina S, Sayre RT. Heavy metal binding properties of Chlamydomonas cells expressing plasmamembrane anchored metallothionein fusion proteins. 2006, (in preparation).
24. Sayre RT, Wagner RE. Method of making microalgal-based animal foodstuff supplements, microalgal-supplemented foodstuffs and method of animal nutrition. US Patent 2005, (Number 6,932,980).
25. Kagi JHR, Schaffer A. Biochemistry of metallothionein. Biochem 1988; 27:8509-8515.
26. Stillman, Martin J. Metallothioneins. Coord Chem Rev 1995; 144:461-511.

27. He Z, Weavers LK, Siripornadulsil S et al. Removal of mercury from sediment by ultrasound combined with Chlamydomonas reinhardtii. Am Chem Soc 2006, (Atlanta, GA).
28. Pinto E, Sigaud-Kutner TCS, Leitao MAS et al. Heavy metal-induced oxidative stress in algae. J Phycol 2003; 39:1008-1018.
29. Hu S, Lau KWK, Wu M. Cadmium sequestration in Chlamydomonas reinhardtii. Pl Sci 2001; 161:987-996.
30. Gekeler W, Grill E, Winnacker EL et al. Algae sequester heavy metals via synthesis of phytochelatin complexes. Archiv Microbiol 1988; 150:197-202.
31. Howe G, Merchant S. Heavy metal-activated synthesis of peptides in Chlamydomonas reinhardtii. Plant Physiol 1992; 98:127-136.
32. Clemens S, Kim EJ, Neumann D et al. Tolerance to toxic metals by a gene family of phytochelatin synthases from plants and yeast. EMBO J 1999; 18:3325-3333.
33. Suk-Bong H, Smith AP, Howden R et al. Phytochelatin synthase genes from Arabidopsis and the yeast Schizosacchromyces pombe. Plant Cell 1999; 11:1153-1163.
34. Vatamaniuk OK, Mari S, Lu YP et al. Mechanism of heavy metal ion activation of phytochelatin (PC) synthase - blocked thiols are sufficient for PC synthase-catalyzed transpeptidation of glutathione and related thiol peptides. J Biol Chem 2000; 275:31451-31459.
35. Lee JG, Ahner BA, Morel FMM. Export of Cadmium and phytochelatin by the marine diatom Thalassiosira weissflogii. Environ Sci Technol 1996; 30:1814-1821.
36. Hu CA, Delauney AJ, Verma DPS. A bifunctional enzyme (Δ^1-pyrroline-5-carboxylate syntetase) catalyzes the first two steps in proline biosynthesis in plants. Proc Nat Acad Sci 1992; 89:9354-9358.
37. Peng Z, Verma DPS. A rice HAL2-like gene encodes a Ca2+-sensitive 3'(2'), 5'-diphosphonucleoside 3'(2')-phosphohydrolase and complements yeast met22 and Escherichia coli cysQ mutations. J Biol Chem 1995; 270:29105-29110.
38. Peng Z, Lu Q, Verma DPS. Reciprocal regulation of Δ^1-pyrroline-5-carboxylate synthetase and proline dehydrogenase genes controls proline levels during and after osmotic stress in plants. Mol Gen Genet 1996; 253:334-341.
39. Siripornadulsil S. Molecular characterization of heavy metal metabolism in transgenic microalgae (Chlamydomonas reinhardtii). Ph D thesis 2002, (Ohio State University).
40. Chaudri AM, Knight BP, Barbosa-Jefferson VL et al. Determination of acute zinc toxicity in pore water from soils previously treated with sewage sludge using bioluminescence assays. Environ Sci Technol 1999; 33:1880-1885.
41. Monciardini P, Podini D, Marmiroli N. Exotic gene expression in transgenic plants as a tool for monitoring environmental pollution. Chemosphere 1998; 37:2761-2772.
42. Mutwakil MHAZ, Reader JP, Holdich DM et al. Use of stress-inducible transgenic nematodes as biomarkers of heavy metal pollution in water samples from an English river system. Archiv Environ Contam Toxicol 1997; 32:146-153.
43. Tauriainen S, Karp M, Chang W et al. Luminescent bacterial sensor for cadmium and lead. Biosens Bioelectron 1998; 13:931-938.
44. Williams RE, Holt PJ, Bruce NC et al. Heavy metals. Biosensors for Environmental Monitoring. 2000:213-225.
45. Rajamani S, Ewalt J, Torres M et al. Transgenic microalgae as heavy metal biosensors. (in preparation)
46. Stryer L, Haugland RP. Energy transfer: A spectroscopic ruler. Proc Natl Acad Sci 1967; 58:719-726.
47. Selvin PR. Fluorescence resonance energy transfer. Methods Enzymol 1995; 246:300-334.
48. Weiss S. Measuring conformational dynamics of biomolecules by single molecule fluorescence spectroscopy. Nat Struct Biol 2000; 7:724-729.
49. Tsien RY. The green fluorescent protein. Annu Rev Biochem 1998; 67:509-544.
50. Cubitt AB, Woollenweber LA, Heim R. Understanding structure-function relationships in the Aequorea victoria green fluorescent protein. Methods Cell Biol 1999; 58:19-30.
51. Truong K, Ikura M. The use of FRET imaging microscopy to detect protein-protein interactions and protein conformational changes in vivo. Curr Opin Struct Biol 2001; 11:573-578.
52. Miyawaki A, Llopis J, Heim R et al. Fluorescent indicators for Ca2+ based on green fluorescent proteins and calmodulin. Nature 1997; 388:882-887.
53. Zaccolo M. Use of chimeric fluorescent proteins and fluorescence resonance energy transfer to monitor cellular responses. Circ Res 2004; 94:866-873.
54. Pearce LL, Gandley RE, Han W et al. Role of metallothionein in nitric oxide signaling as revealed by a green fluorescent fusion protein. Proc Nat Acad Sci 2000; 97:477-482.
55. Nagai T, Ibata K, Park ES et al. A variant of yellow fluorescent protein with fast and efficient maturation for cell-biological applications. Nat Biotechnol 2002; 20:87-90.
56. Zacharias DA, Violin JD, Newton AC et al. Partitioning of lipid-modified monomeric GFPs into membrane microdomains of live cells. Science 2002; 296:913-916.
57. Diener DR, Curry AM, Johnson KA et al. Rescue of a paralyzed-flagella mutant of Chlamydomonas by transformation. Proc Natl Acad Sci 1990; 87:5739-5743.

CHAPTER 10

Hydrogen Fuel Production by Transgenic Microalgae

Anastasios Melis,* Michael Seibert and Maria L. Ghirardi

Abstract

This chapter summarizes the state of art in the field of green algal H_2-production and examines physiological and genetic engineering approaches by which to improve the hydrogen metabolism characteristics of these microalgae. Included in this chapter are emerging topics pertaining to the application of sulfur-nutrient deprivation to attenuate O_2-evolution and to promote H_2-production, as well as the genetic engineering of sulfate uptake through manipulation of a newly reported sulfate permease in the chloroplast of the model green alga *Chlamydomonas reinhardtii*. Application of the green algal hydrogenase assembly genes is examined in efforts to confer H_2-production capacity to other commercially significant unicellular green algae. Engineering a solution to the O_2 sensitivity of the green algal hydrogenase is discussed as an alternative approach to sulfur nutrient deprivation, along with starch accumulation in microalgae for enhanced H_2-production. Lastly, current efforts aiming to optimize light utilization in transgenic microalgae for enhanced H_2-production under mass culture conditions are presented. It is evident that application of genetic engineering technologies and the use of transgenic green algae will improve prospects for commercial exploitation of these photosynthetic micro-organisms in the generation of H_2, a clean and renewable fuel.

Overview

Photosynthesis by unicellular green algae holds promise for generating hydrogen (H_2), a clean and renewable fuel, from nature's most plentiful resources—sunlight and water. Projections of fossil fuel shortfall toward the middle of the 21st century make it important to develop new energy sources that are plentiful, renewable and environmentally friendly. Photobiological H_2 production by green algae meets these requirements. In addition, when scaled-up, it could yield large amounts of cell biomass, a byproduct potentially useful in the food and pharmaceutical industries. This chapter summarizes the state-of-art in this field and offers ideas on how to improve the hydrogen production characteristics of green microalgae through the application of genetic engineering technologies.

Research by Hans Gaffron[1,2] and coworkers over 60 years ago unearthed the ability of unicellular green algae to metabolize H_2. Capable of growth under both anaerobic and aerobic conditions, these photosynthetic eukaryotic microorganisms show remarkable productivity potential. Under ambient mass culture conditions, they typically double their biomass every

*Correspondending Author: Anastasios Melis—University of California, Department of Plant and Microbial Biology, 111 Koshland Hall, Berkeley, California 94720-3102, U.S.A. Email: melis@nature.berkeley.edu

Transgenic Microalgae as Green Cell Factories, edited by Rosa León, Aurora Galván and Emilio Fernández.

24 hours. Perhaps more importantly for the present discussion, the algae also have the ability to generate molecular hydrogen (H_2) photosynthetically, using electrons from water (H_2O).

Light absorption by the photosynthetic apparatus is essential for the generation of H_2, as light facilitates the oxidation of H_2O molecules by photosystem-II (PSII), the release of electrons and protons, and the endergonic transport of these electrons via photosystem-I (PSI) to ferredoxin. The latter is the physiological electron donor to the green algal hydrogenase, whose function is to recombine high potential-energy electrons and protons and to form molecular hydrogen (H_2). Reduced ferredoxin, therefore, links the green algal hydrogenase to the electron transport chain in the chloroplast.[3-5] Oxygen (O_2) is concomitantly evolved as a byproduct in this H_2O-oxidation and electron-transport process of microalgal photosynthesis.

The H_2-production activity of green algae is not manifested in aerobic environments, but must be induced following an obligatory anaerobic incubation of the cells in the dark. Under these conditions, a hydrogenase enzyme is expressed in the chloroplast, the photosynthetic organelle catalyzing light-mediated generation of H_2. In the past, this photosynthetic H_2-production mechanism could function for only short periods of time (30-90 seconds)[6] and generate trace amounts of H_2 that required sophisticated instruments, such as a mass spectrometer or a Clark-type H_2 electrode, for detection. The short life of the process is due to the fact that O_2, evolved as a byproduct of photosynthesis, acts as a powerful inhibitor of the hydrogenase function and a positive suppressor of hydrogenase gene expression. This incompatibility in the simultaneous photo-evolution of O_2 and H_2 has prevented progress in this field for more than half a century.

A recent discovery entailed the development of a novel approach by which to bypass and, therefore, alleviate "the oxygen problem". For the first-time in the 60-years of research in this field, H_2 gas can now be made to accumulate continuously during a modified process of green algal photosynthesis. This breakthrough can be exploited to achieve further improvements in the yield and continuity of hydrogen production.

Sulfur-Nutrient Deprivation Attenuates Photosystem-II Repair and Promotes H_2-Production in Unicellular Green Algae

Photo-oxidative damage occurs frequently within the reaction center of PS II in a light-intensity-dependent manner.[7-10] It causes an irreversible inactivation of the photochemical charge separation and electron transport reactions within the D1 protein of PSII.[7,8] This adverse effect triggers a mechanism leading to the selective removal and replacement of the photoinactivated D1 reaction center protein from the PSII holocomplex (Fig. 1, for a review on this topic please see ref. 11). A temporal and spatial sequence of events concerning the PSII damage and repair cycle has been presented in the literature.[10-14] It is well understood that the PSII damage and repair cycle entails high rates of D1 turnover[15] and requires high rates of de novo D1 biosynthesis in the chloroplast.[16] When PSII repair is impeded, and the inactivated D1 reaction center protein cannot be efficiently replaced, rates of H_2O-oxidation and O_2-evolution in the chloroplast decline.[17] Physiological limitation of the PSII repair process upon sulfur-nutrient deprivation, and a reversible slow-down of D1 replacement, elicited a redirection of chloroplast electron transport in green algae toward H_2-production.[18] It was henceforth realized that genetic engineering of the PSII repair process holds promise for the generation of microalgal strains with constitutive H_2-production properties (see below).

In unicellular green algae, the continuous de novo biosynthesis and replacement of the D1 protein requires a continuous uptake and assimilation of sulfate by the chloroplast.[19] Sulfate assimilation to cysteine and methionine takes place in the chloroplast,[20] and these essential amino acids are required for the subsequent synthesis of proteins in various cellular compartments. The de novo D1 biosynthesis in the chloroplast consumes the majority of cysteine and methionine.[15,16] Therefore, these amino acids must be continuously generated upon sulfate uptake and assimilation, as any surplus amino acids are effectively exported from the chloroplast to the cytosol to serve in the protein biosynthesis needs of this cellular compartment.[19] As

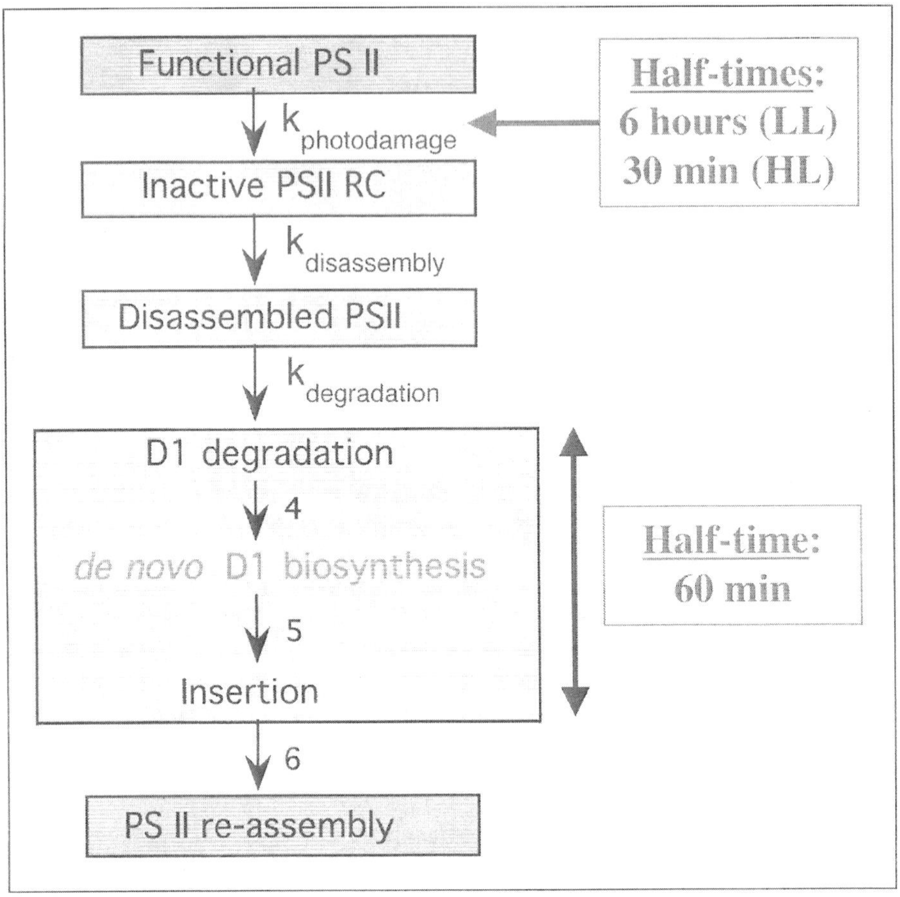

Figure 1. Flow-chart of PSII damage, disassembly, D1 degradation and replacement. The rate of photodamage to PSII (k-photodamage) is directly proportional to the incident light intensity, resulting in variable rates of photodamage, and occurring with halftimes, from 6 hours to 30 minutes. The rate of PSII disassembly (k-disassembly) is not limiting under a broad range of incident intensities. Direct D1 degradation (k-degradation) and de novo D1 biosynthesis (step 4) are required for the repair of PSII from this photo-oxidative damage. The rate of D1 degradation and replacement become severely limiting under sulfur-nutrient deprivation, resulting in loss of photosynthetic water oxidation and oxygen evolution activity.

green algae have no sizable vacuoles in which to store sulfate and other essential inorganic nutrients, it follows that a sulfate deprivation promptly affects rates of de novo D1 protein biosynthesis in the chloroplast.

In the absence of a sufficient supply of sulfur to the chloroplast, D1 protein biosynthesis is impeded and the PSII repair cycle is arrested in the PSII Q_B-nonreducing configuration.[21] Consequently, the rate of photosynthesis declines quasi-exponentially in the light as a function of time under S-deprivation with a half time of about 10-18 h to a mere 5-10% of the control.[18,21-23] This effect is specific to PSII activity in the thylakoid membrane. In brief, interference with the frequent replacement of the D1 PSII reaction center protein by sulfur deprivation causes a decline in the rate of photosynthetic O_2-evolution[18,21] without affecting the rate of mitochondrial respiration.[18,22,24] Within 24 h of S-deprivation, photosynthesis declines to an activity level lower than that of cellular respiration.[18] As respiration dominates the bioenergetic activities of the cell, cultures become anaerobic in the light,[24] leading to the spontaneous expression of the green algal hydrogenase pathway and to H_2-production in the light.[18] It was shown that, under S-deprivation conditions, electrons for the generation of H_2 derived mostly from the residual PSII H_2O-oxidation activity[23-25,28] and in part from endogenous substrate catabolism,[24-28] which feed into the green algal hydrogenase pathway, thereby contributing to the H_2-production process in *Chlamydomonas reinhardtii.* Such physiological attenuation of the photosynthesis-respiration relationship supports continuous H_2 evolution for up to 4 days, and copious amounts of H_2 gas are released from the medium as bubbles via this process. These results were the first experimental demonstration of macroscopic H_2 photoproduction by green microalgae.

Sulfur-nutrient deprivation proved to be a critically successful tool in the sustained production of H_2 by green algae, since for the first time in 60 years of related research, substantial amounts of H_2 were produced continuously for 4-5 days in the light, and H_2 gas could be accumulated in suitable containers. These results suggested that genes and proteins of sulfate metabolism are important in terms of application for enhanced hydrogen production.

Genetic Engineering of Sulfate Uptake in Microalgae for H_2-Production

A recent review article summarized the state-of-the-art in chloroplast sulfate transport in green algae, reporting on genes, proteins and effects of a newly discovered sulfate permease in the model green alga *Chlamydomonas reinhardtii.*[19] Evidence at the molecular genetic, protein and regulatory levels was presented concerning the existence and function of a putative ABC-type chloroplast envelope-localized sulfate permease in *C. reinhardtii.* Four genes, termed *SulP*, *Sulp1*, *Sbp* and *Sabc* were identified, as coding for the transmembrane (SulP and Sulp1) and peripheral (Sbp and Sabc) subunits of this sulfate permease holocomplex.[19] The function of the sulfate permease was probed in antisense transformants of *C. reinhardtii* having lower expression levels of the *SulP* gene. Results showed that chloroplast sulfate uptake capacity was lowered as a consequence of attenuated *SulP* gene expression in the cell, directly affecting rates of de novo protein biosynthesis in the chloroplast.[29] The antisense transformants exhibited phenotypes of sulfate-deprived cells, displaying slow rates of light-saturated O_2 evolution, low levels of Rubisco in the chloroplast and low steady state levels of the PSII D1 reaction center protein. In sealed cultures, such strains become anaerobic (since the activity of photosynthesis is less than that of respiration), show constitutive expression of the green algal hydrogenase pathway and evolve H_2 in the light, even though sulfate nutrients are present in the growth medium. This premise was evidenced in antisense transformants of *C. reinhardtii* cultures, grown in the presence of limiting amounts of sulfate (150 μM), and specifically having lower expression levels of the *SulP* gene.[29,30] Results showed that such *anti-sulP* antisense transformants, but not the wild type, photoproduce H_2 under such conditions.[31] It was concluded that *antisulP* strains are promising as tools to limit the supply of sulfate to the chloroplast genetically, hence leading to continuous H_2-photoproduction in sealed *C. reinhardtii* cultures. This is an interesting new lead for the further development of transgenic microalgae to improve the efficiency, yield, and continuity of the H_2-production process.

Application of the Hydrogenase Assembly Genes in Conferring H_2-Production Capacity in a Variety of Organisms

The application of sulfur deprivation-based methods (see above sections) to manipulate the ratio of photosynthetic O_2 evolution and respiratory O_2 consumption has successfully demonstrated the capacity of *C. reinhardtii* cultures to produce H_2 gas continuously in the light.[31] However, in order to completely uncouple photosynthesis from respiration during H_2-production, it will be necessary to develop organisms that do not depend on respiration to maintain anaerobic conditions. The ideal organism should be able to photo-oxidize water and produce H_2 gas under aerobic conditions, independent of carbon fixation and starch accumulation. The engineering of such an organism requires three main conditions: (a) the green algal hydrogenase gene must be transcribed under aerobic conditions, (b) the catalytic metallo-cluster must be synthesized and assembled into the hydrogenase apoprotein in the presence of O_2, and (c) the mature hydrogenase enzyme must remain functionally active under aerobic conditions.

It has been known that the induction of algal H_2-production activity requires a period of anaerobiosis.[6,32-35] This induction period has been correlated with the accumulation of *HydA1*[36] and *HydA2*[5] gene transcripts, suggesting transcriptional regulation of gene expression by O_2. However, the transcriptional regulation of the green algal hydrogenase genes is a fairly undeveloped area of research. Using the arylsulphatase reporter gene, Stirnberg and Happe[37] identified a DNA region between positions -128 and -21 from the *C. reinhardtii HydA1* transcription initiation site, as required for anaerobic gene expression. Besides O_2, no other factors were assigned as regulators of hydrogenase gene transcription in green algae. However, additional protein factors are required for the proper assembly of the mature enzyme.

The green algal-type hydrogenase catalytic site is defined by the presence of the so-called H-cluster consisting of a 4Fe4S center linked through a cysteine residue to a unique 2Fe2S center. The latter is coordinated by unusual ligands, such as CO and CN, and a purported dithiomethylamine bridge.[38,39] The complex H-cluster structure and the presence of these ligands raised a question as to how the 6Fe-6S metallo-cluster in the catalytic site of the enzyme is incorporated into the apoprotein, as well as how the H-cluster structure is ligated to the hydrogenase apoprotein. Although the maturation of [NiFe]-hydrogenases (a class of mainly H_2-uptake enzymes present in different bacterial genera but not in green algae) is well characterized,[40] very little is known about the maturation of the H-cluster in green algal hydrogenases. In [NiFe]-hydrogenases, the binding of CN and CO to the apoprotein occurs through activation of carbamoyl phosphate to thiocarbamate,[41,42] and the maturation of the metallo-cluster requires 11 different gene products. None of these genes have homologues in unicellular green algae,[43] suggesting that their assembly must occur by a different mechanism. The assembly genes required for the maturation of the H-cluster of green algal hydrogenases, *HydEF* and *HydG*, were only recently discovered in *C. reinhardtii.*[44] These genes are present as three separate ORFs (*HydE, HydF* and *HydG*) in the genome (and in some cases as part of the *Hyd* operon) of all other organisms known to contain similar hydrogenases.[45] They show no homology at all with the [NiFe]-hydrogenase maturation genes. Yet, interestingly, *HydE* and *HydG* display a high degree of homology with the Radical SAM superfamily of proteins.[46,47] Radical SAM proteins are frequently involved in the anaerobic synthesis of complex molecules, and they are known to have labile Fe-S clusters that may function either as catalysts or substrates.[48-51] Interestingly, the NifB protein, which appears to be involved in the assembly of the [FeMo]-cofactor of the nitrogenase enzyme,[52] also belongs to the class of Radical SAM proteins. Recently, the homologous *HydE* and *HydG* proteins from *Thermatoga maritima* were indeed shown to have SAM catalytic activity[53] and to reductively cleave S-adenosyl methionine in the presence of reduced dithionite. *HydF*, on the other hand, contains a GTP-binding site, and it is proposed to function as a GTPase. Transcription and activity of the three green algal hydrogenase maturation proteins are also sensitive to O_2, and require a strictly anaerobic environment. Although the specific role of the three assembly proteins is still not known, on-going efforts in several laboratories are attempting to elucidate the pathways involved in the green algal hydrogenase maturation process. Elucidation of the function of these genes, and of the respective proteins, could find application in commercially exploitable microalgae that are not naturally endowed with the hydrogen production genes and proteins. For example, a transgenic form of *Dunaliella salina*, expressing and assembling the green algal hydrogenase, could find commercial application in hydrogen production.

The first report on the heterologous expression of a green algal hydrogenase (from *Clostridium pasteurianum*) in an organism lacking a native green algal-type hydrogenase (*Synechococcus* PCC7942) was made by Asada and coworkers.[54] Although the apoprotein was successfully expressed, as demonstrated by Northern and Western blot analyses, hydrogenase enzyme activity was not reported. The isolation of the assembly genes required for the maturation of an active hydrogenase,[44,45] opened up the possibility of expressing a variety of green algal-type hydrogenases in *E. coli* for mutagenesis studies aimed at improving characteristics of different enzymes. In this respect, a preliminary report of the heterologous expression of algal hydrogenases in *E. coli* was presented.[44] The expression plasmids, however, were unstable (possibly due to the high GC content of the *C. reinhardtii* genes), which resulted in low levels of hydrogenase expression. Since then, expression systems have been optimized and express now more efficiently algal and a variety of bacterial hydrogenases.[55] The advantages of using *E. coli* as an expression system include (a) growth rates of *E. coli* are much faster than those of *C. reinhardtii*, (b) *E. coli* strains can be optimized for the over-expression of proteins, and (c) gene expression can be regulated by the IPTG promoter. As a consequence, it is possible to easily perform in vitro mutagenesis on hydrogenases, obtain large amounts of the expressed protein in *E. coli*, and purify the expressed proteins via a simple, one-step purification through an affinity column.[55]

Recently, Girbal and coworkers[56] reported the successful expression of a green algal hydrogenase in the anaerobic bacterium, *Clostridium acetobutylicum*, possibly because the organism contains its own native green algal-related hydrogenase and associated maturation proteins homologous to those found in algae. The reported specific activities of H_2 production were higher than those measured by Posewitz et al[44] and by King et al,[55] although the growth and manipulation of *C. acetobutylicum* cultures is more cumbersome than that of *E. coli* with respect to potential future applications. Availability of a versatile system for heterologous expression of the green algal-type hydrogenases in other organisms offers an important new tool to genetically engineer existing hydrogenases that may already display higher tolerance to O_2. These could then be linked to the photosynthetic electron transport chain of green algae for future commercial applications.

Engineering O_2 Tolerance to the Green Algal Hydrogenase

In addition to acting as a suppressor of hydrogenase-related gene transcription, O_2 is also a strong inhibitor of hydrogenase catalytic function. As a result, H_2-production activity is quickly inhibited in the presence of even low concentrations of dissolved O_2.[6,34,35] The mechanism of O_2 inactivation is thought to involve an irreversible oxidation of Fe(II), the distal Fe atom in the 2Fe-2S cluster and the site of proton binding/catalysis.[57] Investigators are currently conducting microorganism surveys in efforts to identify hydrogenases with modified metallo-clusters that are more tolerant to O_2. In the *Ralstonia eutropha* NiFe-hydrogenase, for instance, the presence of an extra CN ligand to Ni causes an increase in the enzyme's O_2 tolerance.[58] Others are addressing the O_2-sensitivity of green algal hydrogenases by preventing O_2 access to the enzyme's catalytic site.[59] Earlier work identified a hydrophobic channel connecting the catalytic H-cluster to the surface of bacterial hydrogenases,[39] and it was proposed that this channel might serve to facilitate access of H_2 to the catalytic site.[60] Similarly-located channels were observed in homology models of the green algal and Clostridial hydrogenases (Kim, unpublished), indicating a conserved function. By using a combination of different computational modeling methods,[61,62] it was demonstrated that H_2 gas molecules may actually diffuse through multiple pathways in the green algal hydrogenase, while O_2 gas is restricted to cavities lining both the hydrophobic channel described above, and a second, previously undetected channel through the protein structure (Fig. 2). Current in vitro and in silico mutagenesis work is being conducted to physically restrict access of O_2 to the active site of the green algal hydrogenase, thus preventing enzyme inactivation under aerobic conditions.

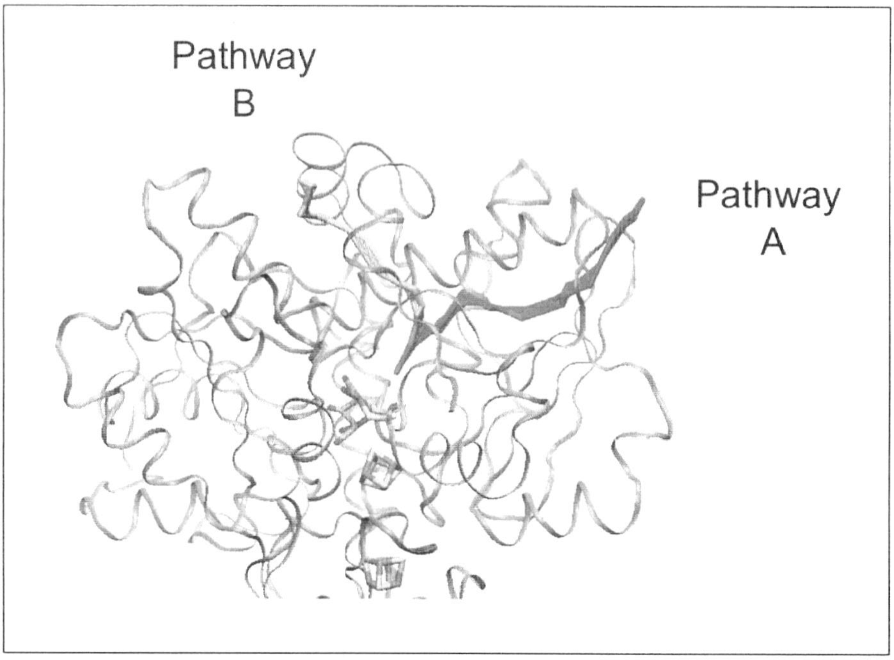

Figure 2. Model of the two O_2-migration pathways through the *Clostridium pasteurianum* CpI hydrogenase, as identified and characterized by the computational techniques described.[62] The O_2-migration pathways are shown in red (Pathway A) and yellow (Pathway B). The iron-sulfur clusters are shown in green and yellow coloration in the lower part of the figure.

Engineering Starch Accumulation in Microalgae for H_2-Production

Sulfur-deprivation induces a prompt Rubisco degradation,[25] a 10-fold starch accumulation in the light[22,25,27,63] and concomitant cell volume enlargement.[25] These substantial morphological and biochemical changes are completed within about 48 h of S-deprivation. In sealed cultures, the starch anabolic activity is followed by a regulated starch degradation and concomitant H_2-photoevolution, which can last for up to 120 h in sulfur-deprived batch cultures.[25] At the end of this period of time, both starch catabolism and H_2-production come to an end. This massive flux of primary metabolites (protein and starch) do not occur in the dark, suggesting a strictly light-dependent and integrated process in metabolite rearrangement and H_2-photoevolution in *C. reinhardtii*.[64] A Rubiscoless and acetate-requiring mutant of *C. reinhardtii* was employed as a tool and compared to the wild type (WT) in terms of protein and starch metabolic flux and H_2-evolution upon sulfur deprivation. The Rubiscoless mutant failed to accumulate starch upon S-deprivation and also failed to evolve and accumulate significant amounts of H_2 gas. These results further strengthened the notion of a carbon substrate requirement for H_2-production under S-deprivation and suggested an obligatory temporal cause-and-effect relationship between the light-dependent catabolism of Rubisco and starch accumulation, and the subsequent ability of the cell to perform a light-dependent starch degradation and H_2-photoevolution.[64] It is apparent that starch catabolism and H_2-production during S-deprivation are intimately linked, although starch degradation could conceivably be substituted for by another carbon source as observed with the starch-less mutant *sta7-10* (see below).

Starch breakdown during H_2-production under S-deprivation conditions is needed to generate the endogenous substrate molecules that feed electrons into the mitochondrial electron transport chain, leading to consumption of O_2 and maintaining anaerobiosis.[4,24] Endogenous substrate catabolism may also feed electrons into the photosynthetic electron transport chain, thereby contributing to photosynthetic H_2 production.[18,23-25,27] It has been shown that H_2 production upon S-deprivation in *C. reinhardtii* lasts so long as there are deposits of starch in the green algal chloroplasts.[25] *C. reinhardtii* mutants aberrant in sugar/starch biosynthesis can grow in the presence of exogenous organic carbon, normally provided in the form of acetate to the culture, but display unusual induction of H_2-production activity under anaerobiosis. This was manifested in the *sta7* mutant of *C. reinhardtii*, isolated by DNA insertional mutagenesis, in which the transforming plasmid interrupted the continuity (hence the function) of an isoamylase gene,[65] encoding a protein responsible for starch accumulation.[66-68] This mutation had a detrimental effect on starch accumulation, photosynthetic hydrogen metabolism, and anaerobic induction of H_2 production in *C. reinhardtii*. This mutant, *sta7-10*, contained only about 3% of insoluble starch compared to WT cells, and was able to induce hydrogenase gene transcription following a short exposure to anaerobiosis. However, hydrogenase gene transcription was entirely suppressed after 7 hours of anaerobic induction in this mutant, and H_2-production activity substantially declined to undetectable levels. Additional starch-less mutants such as *sta6*, defective in the ADP-glucose pyrophosphorylase[69] displayed an even more extreme H_2-production aberrant phenotype.[65] The *sta6* mutant completely lacked starch, had even more attenuated H_2-production activity during the early periods of anaerobic induction, and low levels of hydrogenase gene transcript accumulation.[65] It is clear that other than anaerobiosis, specific metabolite flux conditions are required for the maintenance of hydrogenase gene transcription and H_2-production activity in green algae.

Conversely, recent work by Kruse and coworkers[70] resulted in the isolation of a *C. reinhardtii* strain (*stm6*) with altered respiration and improved H_2-production characteristics. This strain has a modified respiratory metabolism, providing it with two additional important properties: large starch reserves (i.e., enhanced substrate availability), and a low dissolved O_2 concentration (40% of the wild type), resulting in diminished inhibition of hydrogenase induction and prolonged ability to produce H_2 under sulfur-deprivation conditions. These findings strengthen the notion on an interplay between endogenous substrate metabolism and H_2-production.[4,25,27] Starch serves as a cellular endogenous substrate to fuel mitochondrial oxidative phosphorylation during sulfur-deprivation and H_2-production.[4,25,27,63,71] In the particular case of the *stm6* mutant, increased levels of stored starch are likely to be responsible for the enhanced duration of H_2 production observed by Kruse and coworkers.[70]

It is evident from the above achievements that an enhanced starch content in green algae has the potential of enhancing both the duration and the yield of H_2-production in these micro-organisms. These results raise the possibility of genetic engineering approaches to alter the cellular metabolic flux and to induce starch over-accumulation in green algae. For example, this would be achieved upon over-expression of the *Sta7* gene[65] in green algae, resulting in enhancement of the cellular starch content during the course of normal photosynthesis.

Engineering Optimal Light Utilization in Microalgae for H_2-Production

Photosynthesis and H_2-production in unicellular green algae can operate with an absorbed-photon utilization-efficiency that is nearly 100%,[72,73] making these micro-organisms potentially efficient biocatalysts for the generation of H_2 from sunlight and H_2O. Such high photon conversion efficiency is normally manifested under limiting irradiance, e.g., in the shaded ecotype where these green algae normally reside. However, green microalgal culture scale-up and H_2-production would probably take place in photoreactors under direct sunlight, where these unicellular organisms have rather poor photosynthetic quantum efficiencies. The reason for this shortcoming is that, under bright sunlight, the rate of photon absorption by the chlorophyll (Chl) antenna arrays of photosystem II (PSII) and photosystem I (PSI) far-exceeds the rate at which photosynthetic electrons can be utilized. Excess photons cannot be stored in the photosynthetic apparatus but are dissipated (lost) as heat or fluorescence. Up to 80% of absorbed photons could thus be wasted,[74] decreasing light utilization efficiency by the photosynthetic apparatus and resulting in unacceptably low levels of both cellular productivity and H_2 generation. Thus, in a high-density mass culture,[75] cells at the surface would over-absorb and waste sunlight; whereas cells deeper in the culture would be deprived of much needed irradiance (light is strongly attenuated due to filtering by the over layered cells). To attain high photosynthetic performance characteristics in mass culture, it is necessary to minimize the absorption of sunlight by individual cells so as to permit greater transmittance of irradiance through high cell-density cultures. The advent of molecular genetics in combination with sensitive absorbance-difference kinetic spectrophotometry for the precise measurement of the Chl antenna size in green algae now offer a valid approach to reducing the number of photosynthetic Chl antenna molecules.[76]

DNA insertional mutagenesis and screening of the green alga, *C. reinhardtii*, was employed to isolate *tla1*, a stable transformant having a truncated light-harvesting chlorophyll antenna size.[75] Molecular analysis showed a single plasmid insertion into an ORF of the nuclear genome corresponding to a novel gene (*Tla1*) that encodes a protein of 213 amino acids. Biochemical analyses showed the *tla1* mutant to be chlorophyll (but not photosystem) deficient, with the functional chlorophyll antenna size of PSII and PSI being about 50% and 65% of the wild type, respectively. It contained a correspondingly lower amount of light-harvesting proteins than the wild type and had lower steady state levels of *Lhcb* mRNA. The *tla1* strain required a higher light intensity for the saturation of photosynthesis and showed greater solar conversion efficiencies and a higher photosynthetic productivity than the wild type under mass culture conditions.[75] To the best of our knowledge, this is the first identified gene that plays a role in the regulation of the Chl antenna size in oxygenic photosynthesis. The *Tla1* and functionally similar genes, may find direct application in green algal mass culture for biomass accumulation, carbon sequestration and H_2 production. As such, *Tla1* shows promise in helping to overcome the barrier associated with the low light utilization efficiency of photosynthesis during photobiological H_2 production in green algae. In summary, transgenic microalgae with a truncated Chl antenna size are more promising in biotechnological applications that require mass cultivation under direct sunlight than their wild type counterparts.

Future Directions

Photobiological H_2 research has both fundamental and practical value. On the one hand, it addresses the paradox of a light-driven, anaerobic H_2 metabolism in oxygenic unicellular green algae (fundamental research), while on the other, it may open the possibility of a clean, renewable fuel from nature's most abundant resources, sunlight and water (practical value). Green algal H_2 research is still in its infancy, despite its 65-year history, primarily because earlier research could not overcome the mutually exclusive nature of O_2 and H_2 photoproduction in these microorganisms. Recent advances, including (a) demonstration of continuous H_2 photoproduction in green algae as the result of imposing a nutritional deprivation, (b) metabolic engineering of green algae to improve continuity and yield of H_2 production, (c) cloning and sequencing of green algal hydrogenase genes, (d) discovery of the genes required for the maturation of the green algal-type hydrogenases (e) identification of genes that could improve the optical and light absorption properties of the cells, thereby optimizing H_2-production efficiency, are all evidence of progress and manifestation of the current state-of-the-art.

Acknowledgements

The work of the authors discussed in this review was supported in part by the US Department of Agriculture - National Research Initiatives, Competitive Grants Program and the DaimlerChrysler Corporation (A.M.), the US DOE Hydrogen, Fuel Cells and Infrastructure

Technologies Program (A.M., M.S. and MLG), the US DOE Office of Science, Energy Biosciences Program (M.L.G.) and Genomes to Life (M.S.) Programs. M.S. also acknowledges the support of the NREL Director's Discretionary Research and Development Program during preparation of this review article. Figure 2 was kindly generated by Jordi Cohen using VDM software.[77]

References

1. Gaffron H. Reduction of CO_2 with H_2 in green plants. Nature 1939; 143:204-205.
2. Gaffron H, Rubin J. Fermentative and photochemical production of hydrogen in algae. J Gen Physiol 1942; 26:219-240.
3. Florin L, Tsokoglou A, Happe T. A novel type of [Fe]-hydrogenase in the green alga Scenedesmus obliquus is linked to the photosynthetic electron transport chain. J Biol Chem 2001; 276:6125-32.
4. Melis A, Happe T. Hydrogen production: Green algae as a source of energy. Plant Physiol 2001; 127:740-748.
5. Forestier M, King P, Zhang L et al. Expression of two [Fe]-hydrogenases in Chlamydomonas reinhardtii under anaerobic conditions. Eur J Biochem 2003; 270:2750-2758.
6. Ghirardi ML, Togasaki RK, Seibert M. Oxygen sensitivity of algal H_2 production. App Biochem Biotech 1997; 63-65:141-151.
7. Cleland RE, Melis A, Neale PJ. Mechanism of photoinhibition: Photochemical reaction center inactivation in system II of chloroplasts. Photosynth Res 1986; 9:79-88.
8. Demeter S, Neale PJ, Melis A. Photoinhibition: Impairment of the primary charge separation between P680 and pheophytin in photosystem II of chloroplasts. FEBS Lett 1987; 214:370-374.
9. Barber J. Molecular basis of the vulnerability of PSII to damage by light. Aust J Plant Physiol 1994; 22:201-208.
10. Melis A. Photosystem-II damage and repair cycle in chloroplasts: What modulates the rate of photodamage in vivo? Trends in Plant Science 1999; 4:130-135.
11. Yokthongwattana K, Melis A. Photoinhibition and recovery in oxygenic photosynthesis: Mechanism of a photosystem II damage and repair cycle. In: Demmig-Adams B et al, eds. Photoprotection, Photoinhibition, Gene Regulation and Environment. Springer, The Netherlands: 2006:175-191.
12. Melis A. Dynamics of photosynthetic membrane composition and function. Biochim Biophys Acta 1991; 1058:87-106.
13. Aro EM, Virgin I, Andersson B. Photoinhibition of photosystem II. Inactivation, protein damage and turnover. Biochim Biophys Acta 1993; 1143:113-134.
14. Melis A. Photostasis in plants: Mechanisms and regulation. In: Williams TP, Thistle A, eds. Photostasis and Related Phenomena. New York: Plenum Publishing Corporation, 1998:207-221.
15. Mattoo AK, Edelman M. Intramembrane translocation and posttranslational palmitoylation of the chloroplast 32-kD herbicide-binding protein. Proc Natl Acad Sci USA 1987; 84:1497-1501.
16. Vasilikiotis C, Melis A. Photosystem-II reaction center damage and repair cycle - Chloroplast acclimation strategy to irradiance stress. Proc Natl Acad Sci USA 1994; 91:7222-7226.
17. Kim JH, Nemson JA, Melis A. Photosystem-II reaction center damage and repair in the green alga Dunaliella salina: Analysis under physiological and adverse irradiance conditions. Plant Physiol 1993; 103:181-189.
18. Melis A, Zhang L, Forestier M et al. Sustained photobiological hydrogen gas production upon reversible inactivation of oxygen evolution in the green alga Chlamydomonas reinhardtii. Plant Physiol 2000; 122:127-135.
19. Melis A, Chen HC. Chloroplast sulfate transport in green algae: Genes, proteins and effects. Photosynth Res 2005; 86:299-307.
20. Hell R. Molecular physiology of plant sulfur metabolism. Planta 1997; 202:138-148.
21. Wykoff DD, Davies JP, Melis A et al. The regulation of photosynthetic electron-transport during nutrient deprivation in Chlamydomonas reinhardtii. Plant Physiol 1998; 117:129-139.
22. Cao H, Zhang L, Melis A. Bioenergetic and metabolic processes for the survival of sulfur-deprived Dunaliella salina (Chlorophyta). J appl Phycol 2001; 13:25-34.
23. Antal TK, Krendeleva TE, Laurinavichene TV et al. The relationship between photosystem 2 activity and hydrogen production in sulfur-deprived Chlamydomonas reinhardtii cells. Doklady Akad Nauk (Biochemistry and Biophysics) 2001; 381:371-375.
24. Ghirardi ML, Zhang L, Lee JW et al. Microalgae: A green source of renewable H_2. Trends Biotech 2000; 18:506-511.
25. Zhang L, Happe T, Melis A. Biochemical and morphological characterization of sulfur deprived and H_2-producing Chlamydomonas reinhardtii (green algae). Planta 2002; 214:552-561.

26. Kosourov S, Tsygankov A, Seibert M et al. Sustained hydrogen photoproduction by Chlamydomonas reinhardtii: Effects of Culture parameters. Biotech Bioeng 2002; 78:731-740.
27. Kosourov S, Seibert M, Ghirardi ML. Effects of extracellular pH on the metabolic pathways in sulfur-deprived, H_2-producing Chlamydomonas reinhardtii cultures. Plant Cell Physiol 2003; 44:146-155.
28. Antal TK, Krendeleva TE, Rubin AB et al. The dependence of algal H_2 production on photosystem II and O_2-consumption activity in sulfur-deprived Chlamydomonas reinhardtii cells. Biochim Biophys Acta 2003; 1607:153-160.
29. Chen HC, Melis A. Localization and function of SulP, a nuclear-encoded chloroplast sulfate permease in Chlamydomonas reinhardtii. Planta 2004; 220:198-210.
30. Chen HC, Newton AJ, Melis A. Role of SulP, a nuclear-encoded chloroplast sulfate permease, in sulfate transport and H_2 evolution in Chlamydomonas reinhardtii. Photosynth Res 2005; 84:289-296.
31. Melis A. Photosynthetic hydrogen metabolism in unicellular green algae. In: Eaton-Rye J et al, eds. Photosynthesis: A Comprehensive Treatise. Physiology, Biochemistry, Biophysics and Molecular Biology. Vol. 1. The Netherlands: Springer, 2006:175-191.
32. Happe T, Naber JD. Isolation, characterization and N-terminal amino acid sequence of hydrogenase from the green alga Chlamydomonas reinhardtii. Eur J Biochem 1993; 214:475-81.
34. Urbig T, Schulz R, Senger H. Inactivation and reactivation of the hydrogenases of the green-algae Scenedesmus obliquus and Chlamydomonas reinhardtii. Zeitschrift fur Naturforschung 1993; 48:41-45.
35. Happe T, Mosler B, Naber JD. Induction, localization and metal content of hydrogenase in the green alga Chlamydomonas reinhardtii. Eur J Biochem 1994; 222:769-774.
36. Happe T, Kaminski A. Differential regulation of the [Fe]-hydrogenase during anaerobic adaptation in the gree alga Chlamydomonas reinhardtii. Eur J Biochem 2002; 269:1022-1032.
37. Stirnberg M, Happe T. Identification of a cis-acting element controlling anaerobic expression of the HydA gene from Chlamydomonas reinhardtii. In: Miyake J, Igarashi Y, Roegner M, eds. Biohydrogen III. Oxford: Elsevier Science, 2004:117-127.
38. Peters JW, Lanzilotta WN, Lemon BJ et al. X-ray crystal structure of the Fe-only hydrogenase (CpI) from Clostridium pasteurianum to 1.8 Angstrom resolution. Science 1998; 282:1853-1858.
39. Nicolet Y, Piras C, Legrand P et al. Desulfovibrio desulfuricans iron hydrogenase: The structure shows unusual coordination to an active site Fe binuclear center. Structure 1999; 7:13-23.
40. Casalot L, Rousset M. Maturation of the [NiFe] hydrogenases. Trends Microbiol 2001; 9:228-237.
41. Paschos A, Glass RS, Bock A. Carbamoylphosphate requirements for synthesis of the active center of [NiFe]-hydrogenases. FEBS Lett 2001; 488:9-12.
42. Reissmann S, Hochleitner E, Wang H et al. Taming of a poison: Biosynthesis of the NiFe-hydrogenase cyanide ligands. Science 2003; 299:1067-1070.
43. Vignais PM, Billoud B, Meyer J. Classification and phylogeny of hydrogenases. FEMS Micro Rev 2001; 25:455-501.
44. Posewitz MC, King PW, Smolinski SL et al. Discovery of two novel radical SAM proteins required for the assembly of an active [Fe]-hydrogenase. J Biol Chem 2004b; 279:25711-25720.
45. Posewitz MC, King PW, Smolinski SL et al. Identification of genes required for hydrogenase activity in Chlamydomonas reinhardtii. Biochem Soc Trans 2005; 33:102-104.
46. Sofia HJ, Chen G, Hetzler BG et al. Radical SAM, a novel protein superfamily linking unresolved steps in familiar biosynthetic pathways with radical mechanisms: Functional characterization using new analysis and information visualization methods. Nucleic Acids Res 2001; 29:1097-1106.
47. Frey PA. Radical mechanisms of enzymatic catalysis. Annu Rev Biochem 2001; 70:121-148.
48. Ugulava NB, Gibney BR, Jarrett JT. Iron-sulfur cluster interconversions in biotin synthase: Dissociation and reassociation of iron during conversion of [2Fe-2S] to [4Fe-4S] clusters. Biochemistry 2000; 39:5206-5214.
49. Krebs C, Henshaw TF, Cheek J et al. Conversion of 3Fe-4S to 4Fe-4S clusters in native pyruvate formate-lyase activating enzyme: Mössbauer characterization and implications for the mechanism. J Am Chem Soc 2000; 122:12497-12506.
50. Ollagnier-de Choudens S, Sanakis Y, Hewitson KS et al. Reductive cleavage of S-Adenosylmethionine by biotin synthase from Escherichia coli. J Biol Chem 2002; 277:13449-13454.
51. Berkovitch F, Nicolet Y, Wan JT et al. Crystal structure of biotin synthase, an S-Adenosylmethionine-dependent radical enzyme. Science 2004; 303:76-79.
52. Allen RM, Chatterjee R, Ludden PW et al. Incorporation of iron and sulfur from NifB cofactor into the iron-molybdenum cofactor of dinitrogenase. J Biol Chem 1995; 270:26890-26896.
53. Rubach JK, Brazzolotto X, Gaillard J et al. Biochemical characterization of the HydE and HydG iron-only hydrogenase maturation enzymes from Thermatoga maritime. FEBS Lett 2005; 579:5055-5060.

54. Asada Y, Koike Y, Schnackenbert J et al. Heterologous expression of clostridial hydrogenase in the cyanobacterium Synechococcus PCC7942. Biochim Biophys Acta 2000; 1490:269-278.
55. King PW, Posewitz MC, Ghirardi ML et al. Functional studies of the [FeFe] hydrogenase maturation in an Escherichia coli biosynthetic system. J. Bacteriol 2006; 188:2163-2172.
56. Girbal L, von Abendroth G, Winkler M et al. Homologous and Heterologous over expression in Clostridium acetobutylicum and characterization of purified clostridial and algal Fe-only hydrogenases with high specific activities. Appl Environm Microbiol 2005; 71:2777-2781.
57. Adams MWW. The structure and mechanism of iron hydrogenases. Biochim Biophys Acta 1990; 1020:115-145.
58. Bleijlevens B, Buhrke T, van der Linden E et al. The auxiliary protein HypX provides oxygen tolerance to the soluble [NiFe]-hydrogenase of Ralstonia eutropha H16 by way of a cyanide ligand to nickel. J Biol Chem 2004; 279:46686-46691.
59. Ghirardi ML, King PW, Posewitz MC et al. Approaches to developing biological H_2-photoproducing organisms and processes. Biochem Soc Trans 2005; 33:70-72.
60. Nicolet Y, de Lacey AL, Vernede X et al. J Am Chem Soc 2001; 123:1596-1601.
61. Cohen J, Kim K, Posewitz M et al. Molecular dynamics and experimental investigation of H_2 and O_2 diffusion in [FeFe]-hydrogenase. Biochem Soc Trans 2005a; 33:80-82.
62. Cohen J, Kim K, King P et al. Finding gas diffusion pathways in proteins: Application to O_2 and H_2 transport in CpI [FeFe]-hydrogenase and the role of packing defects. Structure 2005b; 13:1321-1329.
63. Tsygankov A, Kosourov S, Seibert M et al. Hydrogen photoproduction under continuous illumination by sulfur-deprived, synchronous Chlamydomonas reinhardtii cultures. J Intern Hydrogen Energy 2002; 27:1239-1244.
64. White AL, Melis A. Biochemistry of hydrogen metabolism in Chlamydomonas reinhardtii wild type and a Rubiscoless mutant. Intl J Hydrogen Energy 2006; 31:455-464.
65. Posewitz MC, Smolinski SL, Kankagiri S et al. Hydrogen photoproduction is attenuated by disruption of an isoamylase gene in Chlamydomonas reinhardtii. Plant Cell 2004a; 16:2151-2163.
66. Mouille G, Maddelein ML, Libessart N et al. Preamylopectin processing: A mandatory step for starch biosynthesis in plants. Plant Cell 1996; 8:1353-1366.
67. Ball S. Regulation of starch biosynthesis. In: Rochaix JD, Goldschmidt-Clermont M, Merchant S, eds. The Molecular Biology of Chloroplasts and Mitochondria in Chlamydomonas. Vol 7. Dordrecht, The Netherlands: Kluwer Academic Publishers, 1998:549-567.
68. Myers AM, Morell MK, James MG et al. Recent progress towards understanding biosynthesis of the amylopectin crystal. Plant Physiol 2000; 122:989-997.
69. Zabawinski C, van den Koornhuyse N, d'Hulst C et al. Starchless mutants of Chlamydomonas reinhardtii lack the small subunit of a heterotetrameric ADP-glucose pyrophosphorylase. J Bacteriol 2001; 183:1069-1077.
70. Kruse O, Rupprecht J, Bader KP et al. Improved photobiological H_2 production in engineered green algal cells. J Biol Chem 2005; 280:34170-34177.
71. Melis A. Green alga hydrogen production: Progress, challenges and prospects. Intl J Hydrogen Energy 2002; 27:1217-1228.
72. Ley AC, Mauzerall DC. Absolute absorption cross sections for photosystem II and the minimum quantum requirement for photosynthesis in Chlorella vulgaris. Biochim Biophys Acta 1982; 680:95-106.
73. Greenbaum E. Energetic efficiency of H_2 photoevolution by algal water-splitting. Biophys J 1988; 54:365-368.
74. Melis A, Neidhardt J, Benemann JR. Dunaliella salina (Chlorophyta) with small chlorophyll antenna sizes exhibit higher photosynthetic productivities and photon use efficiencies than normally pigmented cells. J appl Phycol 1999; 10:515-525.
75. Polle JEW, Kanakagiri S, Melis A. tla1, a DNA insertional transformant of the green alga Chlamydomonas reinhardtii with a truncated light-harvesting chlorophyll antenna size. Planta 2003; 217:49-59.
76. Melis A. Bioengineering of green algae to enhance photosynthesis and hydrogen production. Chapter 12. In: Collins AF, Critchley C, eds. Artificial Photosynthesis: From Basic Biology to Industrial Application. Wiley-Verlag and Co., 2005:229-240.
77. Humphrey W, Dalke D, Schulten K. VDM—visual molecular dynamics. J Molec Graphics 1996; 14:33-38.

Chapter 11

Microalgal Vaccines

Surasak Siripornadulsil, Konrad Dabrowski and Richard Sayre*

Abstract

A variety of recombinant vaccines and vaccine delivery systems are currently under development as alternatives to vaccines produced in animals that are primarily administered by injections. These nonanimal alternatives do not transmit animal pathogens, are often rapid to develop, and can be produced on a large scale at low costs. Many of these new vaccine technologies are based on oral delivery systems and avoid the risks of disease transmission associated with the use of syringes for injectable vaccines. In addition, many of these novel systems have extended shelf life, often not requiring refrigeration and thus are applicable in developing countries or remote locations. Here we describe the development of microalgal-based immunization systems. Antigens expressed in the chloroplast or anchored to the surface of plasma membrane are shown to effectively immunize fish and rabbits. The effective oral delivery of antigens by microalgae provides a safe and inexpensive mechanism to immunize animals. The applications of microalgal vaccines are currently being investigated.

Introduction

Disease is a major productivity constraint of intensive agricultural systems. This is particularly true for high-density farm operations. Some of the highest animal densities occur in aquaculture systems. In addition, the host (fish) and the pathogen occupy the same environment. A consequence of these potential risks is that fish farms need to be constantly monitored for disease outbreaks. Containment of a disease outbreak may require sacrificing the entire production run if the outbreak cannot be controlled. Furthermore, disease outbreaks in aquaculture systems may be transmitted to wild-fish populations.

Disease control measures include prevention, selection for disease resistant strains, pre or post-infection treatment with antibiotics, and vaccination. As discussed below, each of these strategies has inherent advantages and limitations.

- Disease prevention strategies such as disinfection, low stress environments, bio-control and bio-containment are cost-effective until an infection occurs.[1,2] Once infection has occurred in a high-density farming operation, the disease often spreads rapidly and other control options must be employed to prevent the spread of the pathogen.
- Selection for disease-resistant strains can be effective, but to date, has not proven effective for many species.
- Antibiotics can treat many bacterial diseases but are not effective against viral and many parasitic diseases.[3] Antibiotic treatment also typically occurs after infection which may be too late or require fast diagnosis before losses occur. In addition, many sick animals have poor appetite, which compromises the effective delivery of antibiotics via feed. An additional concern is the over use and release of antibiotics into the environment. As much as

*Corresponding Author: Richard Sayre—Department of Plant Cellular and Molecular Biology, Ohio State University, Columbus, Ohio, 43210, U.S.A. Email: sayre.2@osu.edu

Transgenic Microalgae as Green Cell Factories, edited by Rosa León, Aurora Galván and Emilio Fernández.

80% of the antibiotic passes through the fish gut. The excretion of large amounts of antibiotics into the environment can result in selection for antibiotic resistant pathogens that exacerbates the disease problem.[3,4] Finally, the presence of antibiotics in food is an issue of major concern for consumers and governmental regulatory agencies. The proliferation of antibiotic-resistant human pathogens has been associated with the wide spread use of antibiotics in agriculture.

- Vaccination is often the most cost-effective strategy for controlling disease (bacterial, fungal and viral). Vaccines are prophylactic, have minimal side effects and reduce environmental pollution. Vaccines have proven effective in controlling a number of diseases in fish.[5-7]

Oral Vaccines

Currently, most vaccines are injected into fish. Vaccination by injection, however, has many liabilities. It is labor intensive, costly, stressful to animals, may cause tissue adhesions (resulting in reduced growth rates and poor quality meats and reduced consumer acceptance), and cannot be done with very young fish (larval-juvenile size). An alternative to injectable vaccines is oral vaccination. Oral vaccine delivery avoids the stresses associated with injectable vaccines and substantially reduces labor costs (Table 1).

There are several potential limitations for oral vaccines including cost, efficacy against some antigens and the potential for limited humoral expression of antibodies. Given these constraints, however, the advantages of oral delivery versus injection clearly indicate that oral vaccination is the preferred delivery approach. Presently, most oral vaccines are delivered to fish in a microencapsulated form to protect antigens from degradation.[6,8-10] Antigens are encapsulated in polysaccharide coated beads 1-10 microns in diameter. The encapsulation ensures that the antigen passes through the acid environment of the stomach so that the antigens may be endocytosed in the posterior intestine.[11]

Microalgal Vaccines

Recently, a novel vaccine delivery system has been developed using the single-celled alga, *Chlamydomonas reinhardtii.*[11] Microalgae, and in particular *Chlamydomonas*, have many features that are desirable for vaccine delivery systems.

- *Chlamydomonas* is amenable to genetic manipulation. This means that foreign genes and proteins (antigens) can be readily expressed in the chloroplast (see Chapters 4 and 8 for

Table 1. Comparative advantages and disadvantages of different vaccine delivery systems

Injectable	Oral/Immersion
Effective against a broad range of pathogens.	May be more effective against some pathogens than others.
Stressful to fish.	No stress to fish.
Not useable for fish less than several centimeters in length.	Effective for all stages of fish once immune system is competent.
May induce both humoral and mucosal immune response.	More effective in inducing cellular, mucosal (intestinal) or epithelial (skin) immune response than humoral response.
High labor or equipment costs.	Minimal expense involved in delivery of vaccine.
May induce tissue adhesions resulting in reduced growth rates.	No tissue adhesions.

more details), mitochondria (prokaryotic-like expression system), or the cytoplasm (eukaryotic-like expression system) with high yields ($\geq$1% of total cellular protein).[11]

- Algal vaccines are inexpensive. Protein production levels as high as 2-5 mg protein/liter of cells have been achieved at a relatively low cost (<$1/mg protein). In contrast, the cost of producing a synthetic peptide antigen ranges between $35 and $95/mg peptide. In addition, microalgal produced antigens do not need to be isolated and purified from any other source; the delivery system (algae) produces the antigen.
- Algae are a potential food source for larval fish and are the proper size (10 μm) for direct uptake by many species.[8]
- *Chlamydomonas* is innocuous, nontoxic and nonpathogenic. Microalgae also do not harbor animal pathogens.
- Both fresh water (*C. reinhardtii*) and marine species (*C. pulsatilla*) of *Chlamydomonas* are available.
- *Chlamydomonas* mutants are available that preclude growth and mating outside of a controlled environment. In addition, transgenic microalgae can be grown in biological containment.
- Algae are compatible with feed processing technologies and are less likely to be confused with other feed ingredients, thus limiting the potential for accidentally contaminating feeds with a transgenic organism.

Recent Progress

To determine if it was feasible to use microalgae as a vaccine delivery system a highly immunogenic protein was chosen for expression in transgenic *Chlamydomonas*. The first antigen chosen for testing using the micoalgal antigen delivery system was the p57 antigen, the causative agent of bacterial kidney disease (BKD). BKD is caused by the intracellular bacterium *Renibacterium salmoninarum* and is the most important disease of wild and farmed salmonids from an economic point of view.[12-17] BKD is characterized by the presence of lesions around the eyes, swollen abdomen, blood-filled blisters, and ulcers and lesions in the kidney, liver, heart and spleen.[12-17] Significantly, symptoms typically take several weeks to develop following infection.[12-17] By the time BKD symptoms develop it is often too late to administer antibiotics to control the disease in infected fish.

We tested the BKD p57 antigen in fish for several reasons. First, one of the hallmarks of BKD pathobiology is the production and secretion of the p57 leukocyte agglutinating protein. The p57 protein is highly antigenic and therefore a good target to determine whether the microalgal delivery of an antigen could induce an immune response in vaccinated animals.[18,19] Unfortunately, while the p57 protein is highly immunogenic, it has not been shown to be a very effective (single) vaccine target. This is presumably due to its secretion by *R. salmoninarum*.[18-25]

Two different strategies were used to express p57 antigens in *Chlamydomonas*, (1) as a p57 antigenic domain (14 amino acids) fused to a plasma membrane protein (Fig. 1, hereafter referred to as the E-22 fusion protein), and (2) as the intact holo-protein expressed in the chloroplast (hereafter referred to as CP57). Algae expressing the E-22 fusion protein were generated by electroporation using the *Chlamydomonas* strain CC-425 (an arginine auxotrophic, cell wall less strain) as a host. Algae expressing the CP57 protein were generated by particle bombardment using strain CC-744 (*psbA* deletion strain) as the host.[26] Expression of the chloroplast-encoded CP57 protein and the nuclear encoded E-22 fusion protein were confirmed by western blot using mouse anti-p57 monoclonal antibodies (data not shown).

The primary concern as to whether the *Chlamydomonas* vaccination system would effectively induce an immune response in fish was whether the intact antigen could be delivered to the posterior intestine-associated, antigen presenting cells. The major issues were whether the antigen would pass intact through the stomach and whether the antigen would be taken up by gut-associated phagocytotic cells (GAPC). Significantly, wild-type *Chlamydomonas* has a very

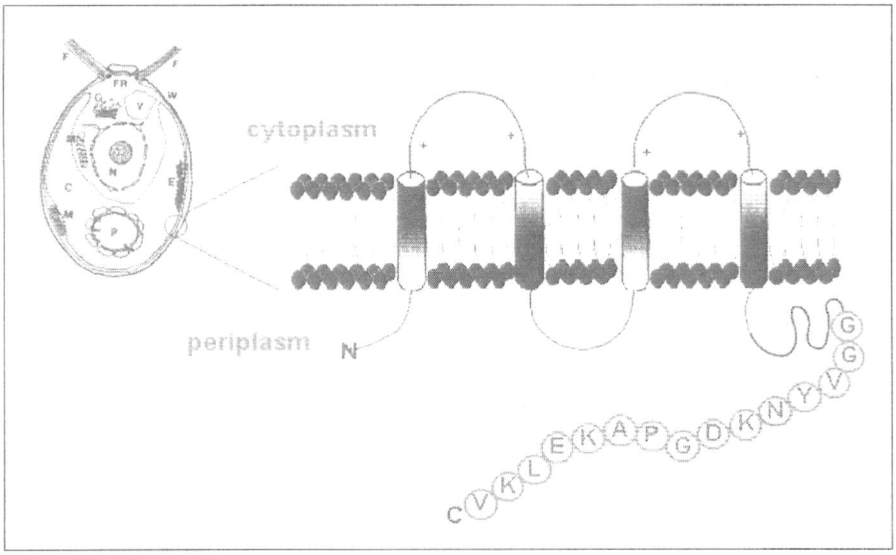

Figure 1. Periplasmic expression of a gene fusion between a p57 antigenic determinant and the high CO_2 induced membrane protein of *Chlamydomonas.*

rigid cell wall composed of glycosylated and sulfated hydroxyproline-rich proteins.[27] These cell walls can tolerate mechanical shear forces up to several thousand pounds/inch^2 before rupture. Furthermore, the cell wall is resistant to acid treatments. In theses studies both walled and wall-less cells were used to deliver antigens. In addition, *Chlamydomonas* cells have an optimal size (1 - 10 μm diameter) for phagocytosis by gut-associated phagocytes.[28-30] Finally, the peptidoglycan cell wall of *Chlamydomonas* may have adjuvant-like properties similar to bacterial cell walls facilitating potentiation of the immune system.

To determine whether microalgal vaccines would induce an immune response, juvenile rainbow trout were either immersed in antigen-expressing algae or were fed diet containing freeze-dried algae and subsequently monitored for induction of an antigen-specific immune response. As shown in the immunoblot below (Fig. 2), fish fed algae (4% algal dry weight of feed, fed to satiation) expressing the E-22 fusion protein or the CP57 protein produced circulating antibodies that recognized the p57 protein expressed in *Chlamydomonas.* Significantly, juvenile trout fed diet containing wild-type algae or no algae, did not generate antibodies against the p57 protein. Identical results were achieved when trout were immersed (two hours with 10^6 cells/mL) in live transgenic algae expressing the p57 antigens. In contrast to the results achieved by feeding freeze-dried algae, fish that were immersed in live algae produced p57-specific immunoglobulins (IgM) only in the mucus (Fig. 2, data not shown). These results demonstrate that microalgal delivery of antigens by feed or by immersion can induce antibody production in different tissues (blood or skin epithelial cells, respectively). Since most pathogens generally enter the host through epithelial tissues, these results suggest that the most effective means to immunize fish with microalgal vaccines is to provide the immunogen both in the feed and by immersion.

To determine if the range of animals that could be immunized using microalgal antigen delivery systems could be extended, rabbits were also immunized with transgenic algae expressing the CP57 antigen in the chloroplast. Rabbits were fed either freeze-dried algae incorporated into the feed pellet by cold extrusion or were allowed to drink live algae expressing the antigen. As shown on western blots (Fig. 3) a p57-specific antibody was generated in

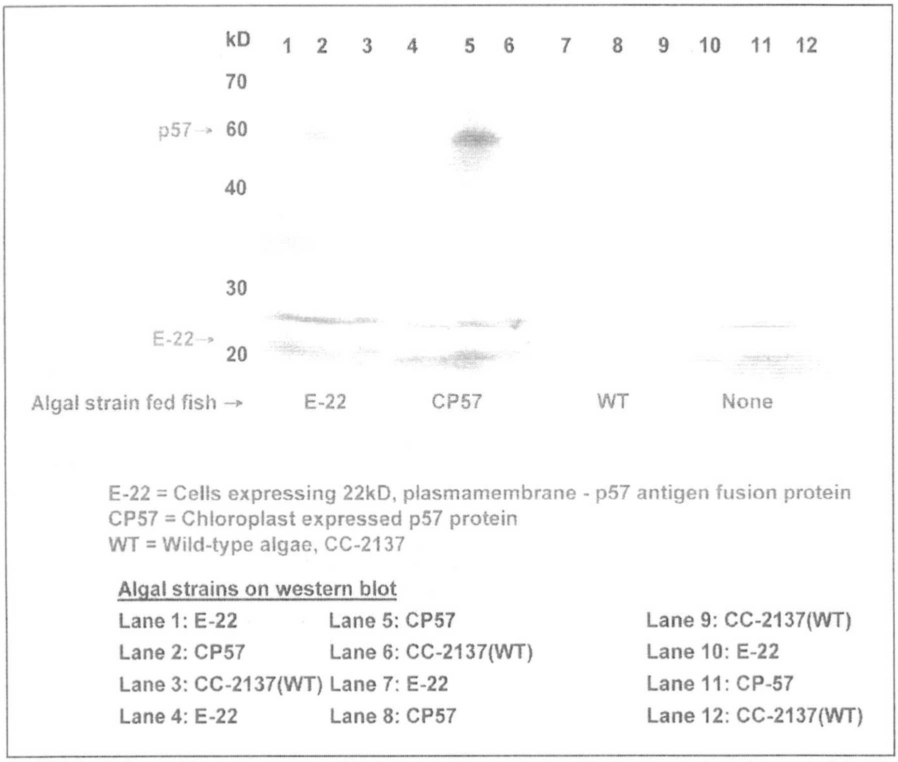

Figure 2. Western blot of wild-type and transgenic microalgae expressing the intact p57 antigen (CP57) in the chloroplast or the p57 antigenic epitope fused to a plasma membrane protein (E-22) probed with sera from fish fed feed lacking algae, feed containing wild-type (WT) or feed containing transgenic algae expressing p57 antigens (CP57 or E-22). Identical results were achieved using mucus obtained from juvenile trout immersed in either no algae, live wild-type algae or live transgenic algae expressing the p57 antigens (data not shown).

rabbits immunized with live microalgae expressing the p57 antigen in the chloroplast (*psbA* deletion strain CC744 as host). No p57 cross-reacting antibodies were detected in preimmunized rabbits. In addition, to detecting the p57 protein expressed in algal chloroplast, sera from CP57 immunized rabbits detected p57 protein in transgenic microalgae expressing the intact p57 protein in the cytoplasm. Significantly, no antibodies were detected from sera collected from animals fed an equivalent mass of freeze-dried microalgae expressing the p57 protein as a dry feed. These results indicate that only live intact algae expressing the p57 antigen could apparently induce a p57-specific immune response in rabbits. The overall cost of the microalgal vaccine dose used to immunize the rabbits was \$0.001/ animal using laboratory grown algae.

The potential of microalgal vaccines remains to be determined. Issues that remain to be addressed include enhancing the levels of antigen expression and display in microalgae, the effects of post-translational modifications (glycosylation) on antigen immunogenicity, and identifying the range of animals that can be effectively immunized using microalgal vaccines. Ultimately, the outcomes of pathogen challenge and survival trials of microalgal vaccinated animals will determine whether microalgal vaccines offer an effective, inexpensive and safe means to vaccinate animals.

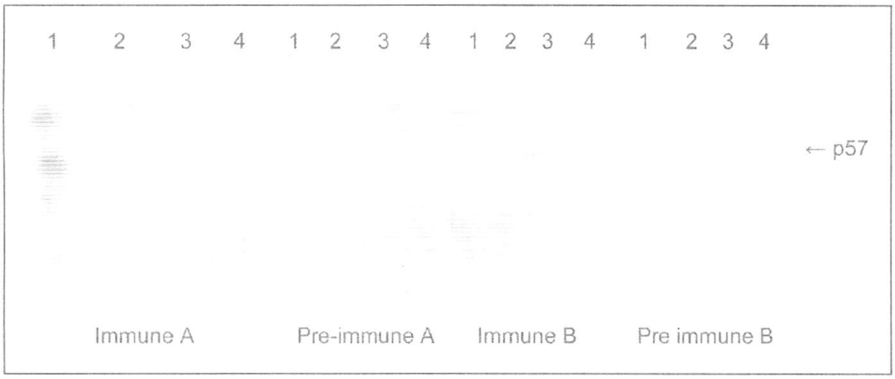

Figure 3. Western blot of wild-type and transgenic microalgae expressing the intact p57 antigen (NP57) in the cytoplasm or the chloroplast (CP57) probed with sera from preimmunized rabbits or immunized rabbits fed wild-type (WT) algae or transgenic algae expressing the CP57 antigen. Lane 1, wild-type, cell wall-less strain, CC425; lane 2, transgenic algae expressing the NP57 antigen in a CC425 background; lane 3, transgenic algae expressing the CP57 antigen in a CC744 background (*psbA* deletion mutant); lane 4 wild-type strain, CC744. The location of the p57 protein is indicated by the arrow. Rabbits were given live cells (100 mL, 10^8 cells total) equivalent to 0.02 gDW^{-1} as a primary and as a secondary (boost) immunization three weeks apart. Sera were collected at week five. Sera were preincubated with WT algal cell extracts to reduce interactions with algal proteins and used on the western blot at a 1:400 titer. Equivalent cell numbers of algae were loaded in each lane.

Acknowledgements

We thank Dr. Richard Wagner for Dr. Mary Ann Abiado for technical contributions (fish survival in algae and editorial comments, respectively). This research was funded by the National Oceanographic and Atmospheric Administration, Ohio Sea Grant Program (RTS) and the Ohio Agricultural Research and Development Center (KD).

References

1. Lavilla-Pitogo CR, Leano EM, Paner MG. Mortalities of pond-cultured juvenile shrimp, Penaeus monodon, associated with dominance of luminescent vibrios in the rearing environment. Aquaculture 1998; 164:337-349.
2. Moriarty DJW. Control of luminous Vibrio species in penaeid aquaculture ponds. Aquaculture 1998; 164:351-358.
3. Park ED, Lightner DV, Park DL. Antimicrobials in shrimp aquaculture in the United States: Regulatory status and safety concerns. Rev Environ Contam Toxicol 1994; 138:1-20.
4. Pothuluri JV, Nawaz MS, Khan AA et al. Antimicrobial use in aquaculture: Environmental fate and potential for transfer of bacterial resistance genes. Recent Res Dev Microbiol 1998; 2:351-372.
5. Sindermann CJ. Principal Diseases of Marine Fish and Shellfish, Vol. 2. 2nd ed. San Diego: Academic Press Inc., 1990.
6. Uchida T, Goto. Oral delivery of poly (DL-lactide-coglycolide microspheres containing ovalbumin as vaccine formulation: Particle size study. Biol Pharm Bull 1994; 17:1272-1276.
7. Walmsley AM, Arntzen CJ. Plants for delivery of edible vaccines. Curr Opinion Biotechnol 2000; 11:126-129.
8. Ellis AE, ed. Fish Vaccination. Berkeley: Academic Press, 1998.
9. Lavelle EC, Jenkins PG, Harris JE. Oral immunization of rainbow trout with antigen microencapsulated in poly (DL-lactide-coglycolide) microparticles. Vaccine 1997; 15:1070-1078.
10. Rombout JHWM, Lamers CHJ, Helfrich MH et al. Uptake and transport of intact macromolecules in the intestinal epithelium of carp (Cyprinus carpio L.) and the possible immunological implications. Cell Tissue Res 1985; 239:519-30.

11. Georgopoulou U, Dabrowski K, Sire MF et al. Absorption of intact proteins by the intestinal epithelium of trout, Salmo-Gairdneri - A luminescence enzyme immunoassay and cytochemical study. Cell Tissue Res 1988; 251:145-152.
12. Sayre RT, Wagner RE, Siripornadulsil S et al. Use of Chlamydomonas reinhardtii and other transgenic algae in food or feed for delivery of antigens. 2001, (WO 2001098335, A2 20011227).
13. Barton TA, Bannister LA, Griffiths SG et al. Further characterization of Renibacterium salmoninarum extracellular products. Applied Environ Microbiol 1997; 63:3770-3775.
14. Austin B, Rayment JN. Epizootiology of Renibacterium salmoninarum, the causal agent of bacterial kidney disease in salmonid fish. J Fish Dis 1985; 8:505-509.
15. Austin B, Austin DA. Bacterial Fish Pathogens: Disease in Farmed and Wild Fish. Chichester, UK: Ellis Horwood Ltd., 1987:364.
16. Bruno DW, Munro ALS. Observations on Renibacterium salmoninarum and the salmonid egg. Dis Aquat Org 1986; 1:83-87.
17. Bullock GL, Herman RL. Bacterial kidney disease of salmonids fishes caused by Renibacterium salmoninarum. Fish Disease Leaflet 1986; 78, (<http://www.lsc.nbs.gov/fhl/fdl/fdl78.htm>).
18. Sanders JE, Pilcher KS, Fryer JL. Relation of water temperature to bacterial kidney disease in coho salmon (Oncorhynchus kisutch), sockeye salmon (O. nerka), and steelhead trout (Salmo gairdneri). J Fish Res Board Can 1978; 35:811-820.
19. Barton TA, Bannister LA, Griffiths SG et al. Further characterization of Renibacterium salmoninarum extracellular products. Applied Environ Microbiol 1997; 63:3770-3775.
20. Chien MS, Gilbert T, Huang C et al. Molecular cloning and sequence analysis of the gene coding for the 57-kDa major soluble antigen of the salmonid fish pathogen Renibacterium salmoninarum. FEMS Microbiol Lett 1992; 96:259-66.
21. Evenden AJ, Grayson TH, Gilpin ML et al. Renibacterium salmoninarum and bacterial kidney disease-the unfinished jigsaw. Ann Rev Fish Dis 1993; 87-104.
22. Fredriksen A, Endresen C, Wergeland HI. Immunosuppressive effect of a low molecular weight surface protein from Renibacterium salmoninarum on lymphocytes from Atlantic salmon (Salmo salar L.). Fish Shellfish Immunol 1997; 7:273-282.
23. Gómez-Chiarri M, Brown LL, Levine RP. Protection against Renibacterium salmoninarum infection by DNA-based immunization. Aquaculture Biotechnology Symposium Proceedings. Physiology Section of the American Fisheries Society, 1996:155-157.
24. Griffiths SG, Melville KJ, Salonius K. Reduction of Renibacterium salmoninarum culture activity in Atlantic salmon following vaccination with avirulent strains. Fish Shellfish Immun 1998; 8:607-619.
25. Kaattari SL, Piganelli JD. Immunization with bacterial antigens: Bacterial kidney disease. Dev Biol Stand 1997; 90:145-152.
26. Wood PA, Kaattari SL. Enhanced immunogenicity of Renibacterium salmoninarum in chinook salmon after removal of the bacterial cell surface-associated 57 kDa protein. Dis Aquat Org 1996; 25:71-9.
27. Ruffle SV, Sayre RT. Functional analysis of photosystem II. In: Goldshmidt-Cleremont M, Merchant S, Rochaix JD, eds. Molecular Biology of Chlamydomonas: Chloroplasts and Mitochondria, Chapter 16. Kluwer Academic Publishers, 1998.
28. Harris EH. The Chlamydomonas Sourcebook: A Comprehensive Guide to Biology and Laboratory Use. San Diego: Academic Press, 1989.
29. Gudding R, Lillehaug A, Evensen. Recent developments in fish vaccinology. Vet Immunol Immunopathol 1999; 72:203-212.
30. Newman SG. Bacterial vaccines for fish. Ann Rev Fish Dis 1993; 3:145-85.

Index

S

T

U

W

X

Y